Advances in Intelligent Systems and Computing

Volume 257

Series Editor

Janusz Kacprzyk, Polish Academy of Sciences, Warsaw, Poland
e-mail: kacprzyk@ibspan.waw.pl

For further volumes:
http://www.springer.com/series/11156

About this Series

The series "Advances in Intelligent Systems and Computing" contains publications on theory, applications, and design methods of Intelligent Systems and Intelligent Computing. Virtually all disciplines such as engineering, natural sciences, computer and information science, ICT, economics, business, e-commerce, environment, healthcare, life science are covered. The list of topics spans all the areas of modern intelligent systems and computing.

The publications within "Advances in Intelligent Systems and Computing" are primarily textbooks and proceedings of important conferences, symposia and congresses. They cover significant recent developments in the field, both of a foundational and applicable character. An important characteristic feature of the series is the short publication time and world-wide distribution. This permits a rapid and broad dissemination of research results.

Maria Mach-Król · Tomasz Pełech-Pilichowski
Editors

Advances in Business ICT

 Springer

Editors
Maria Mach-Król
University of Economics in Katowice
Katowice
Poland

Tomasz Pełech-Pilichowski
AGH University of Science and Technology
Krakow
Poland

ISSN 2194-5357   ISSN 2194-5365   (electronic)
ISBN 978-3-319-03676-2   ISBN 978-3-319-03677-9   (eBook)
DOI 10.1007/978-3-319-03677-9
Springer Cham Heidelberg New York Dordrecht London

Library of Congress Control Number: 2013954679

Printed on acid-free paper

Springer is part of Springer Science+Business Media (www.springer.com)

Editors' Foreword

We are witnessing more and wider use of ICT technologies, in particular for business purposes. Software and hardware solutions based on simple data processing and visualization don't provide capabilities for advanced data analysis, decision supporting and processing of large data sets aimed at extraction of relevant information. This gave rise to researchers' and business sector's interest in solutions for processing and acquisition of information.

Futurists and scientists alike profess the coming of a new era in the history – the knowledge era. The notion of knowledge is as old as humans' self-consciousness, but new challenges appear. The meaning of the word "knowledge" is changing from cognitive notion to a technical term denoting a structured economic resource to be actively managed.

The same process of change applies to the notion of intelligence. Nowadays the feature of being "intelligent" may be attributed not only to humans, but also to computer systems. Therefore it is not surprising that intelligent systems may be used to actively manage knowledge in an enterprise. And one of the best answers from computer scientists to knowledge managers is the use the Business ICT, such as Business Intelligence, reasoning systems, advanced technologies of data processing, content processing and information indexing, knowledge management for better decision support, collaboration and competitiveness, and may others.

This contributed volume is a result of vivid and extremely valuable discussions held at 3rd International Workshop on Advances in Business ICT (ABICT) in Wrocław, Poland, September 9-12, 2012. The workshop focused on Advances in Business ICT approached from a multidisciplinary perspective. It provided an international forum for scientists/experts from academia and industry to discuss and exchange current results, applications, new ideas of ongoing research and experience on all aspects of Business Intelligence. ABICT has also been an opportunity to demonstrate different ideas and tools for developing and supporting organizational creativity, as well as advances in decision support systems.

This book is of interest to researchers, widely understood business, public sector and IT professionals. It consists of eight chapters which present a broad spectrum of research results on business intelligence systems design and implementation, business processes modeling, business rules description languages, problems of data integration from enterprise data warehouses, performance issues of simulation models, possibilities of using temporal logics for knowledge management and problems of legal information digitalization and legal text processing.

Maria Mach-Król
Tomasz Pełech-Pilichowski

Contents

Simulation Driven Development of the German Toll System – Simulation Performance at the Kernel and Application Level

Tommy Baumann, Bernd Pfitzinger, and Thomas Jestädt

Abstract. Simulation driven development – the idea of using simulation models as executable system specification in any phase of the system development process [4] – depends on the performance of the simulation model and execution framework. We study the performance issues of an existing large-scale simulation model of the German toll system using a discrete-event simulation (DES) model. The article first introduces the German toll system and the simulation framework developed to analyze the systems' behavior. To address the simulation performance the article describes a number of common performance limitations of several commercial and non-commercial DES simulation kernels. These performance limitations are addressed in kernel-level benchmarks. At the application-level a DES implementation of the German toll system is used to compare two commercial DES tools and several optimizations are introduced both on the simulation model and kernel level to achieve the necessary performance for a detailed and realistic simulation of a fleet of 750 000 trucks.

Tommy Baumann
Andato GmbH & Co. KG, Ehrenbergstraße 11, 98693 Ilmenau, Germany
e-mail: tommy.baumann@andato.com

Bernd Pfitzinger
Toll Collect GmbH, Linkstraße 4, 10785 Berlin, Germany
FOM Hochschule für Oekonomie & Management, Bismarckstraße 107,
10625 Berlin, Germany
e-mail: bernd.pfitzinger@toll-collect.de

Thomas Jestädt
Toll Collect GmbH, Linkstraße 4, 10785 Berlin, Germany
e-mail: thomas.jestaedt@toll-collect.de

M. Mach-Król and T. Pełech-Pilichowski (eds.), *Advances in Business ICT*,
Advances in Intelligent Systems and Computing 257,
DOI: 10.1007/978-3-319-03677-9_1, © Springer International Publishing Switzerland 2014

1 Introduction

Software evolution is a fact of life. Software-intensive systems become ever larger and to make matters worse include ever more distributed endpoints up to mobile and ubiquitous computing [11]. Introducing changes and new features to an existing system is both time consuming and error prone – one study [17] claims that the probability of critical problems due to poor design decisions is over 60% in the specification phase. Simulations are a vital step in the design of systems or the assessment of planned changes [3, 4] – reducing the inherent risk of ongoing system development and allowing for a faster system deployment. In addition simulations predict the dynamic system behavior which can become highly non-linear or chaotic even for simple systems [25].

To specify and evaluate the German toll system, a simulation-driven design approach has been selected [5]. The approach is characterized by applying modeling and simulation technologies in the early design stages, i.e. at at a time when most of the important design decisions have to be made. As a result both the systems and processes are specified in the form of executable models. The approach allows to validate and optimize the overall system architecture already in the specification phase – avoiding expensive integration issues in the subsequent implementation and integration phases.

Consequently, specification speed and quality is considerably increased while the system and product uncertainty is decreased. It is noteworthy that simulation-driven design not only refers to the system under design but also includes the surrounding design process, i.e. the process is also captured as an executable specification which allows automating design steps like architecture optimization, validation against operational scenarios and tracking of design decisions.

A prerequisite to applying executable models is a so called execution domain: In our context Discrete Event Simulation (DES, [19]) has gained significance. We choose DES as the execution domain of our simulation model (although in future work the behavior of the user interaction might better be modeled in an agent-based approach). DES is used in many industries, e.g. energy, telecommunications, production, logistics, avionics, automotive, business processes and system design. Inter alia DES is applied for dimensioning of resources, to answer questions about topology, scalability and performance regarding operational scenarios, to predict system behavior and to estimate risks.

Increasingly the performance in defining and executing models becomes vital due to the increased complexity of systems and processes as well as the customer requirement to create holistic, integrated, high accuracy models up to real world scale. Several use cases of simulations are only possible once the simulation performance is 'good enough': simulating the longterm dynamic behavior, iterative optimization loops, automatic test batteries, real-time models (higher reactivity to market demands and changes) and automated specification and modeling processes (including model transformation/generation) [28].

The outline of the article is as follows: Section 2 gives an overview of the automatic German toll system, the corresponding simulation model and typical simulation results. Section 3 introduces the performance properties of discrete event simulations and discusses appropriate performance metrics and benchmarking models. This is followed in section 4 by a discussion of the simulation performance and scaling of several DES tools for basic simulation operations. Using an existing microscopic holistic execution specification of the automatic German toll system [27] we describe several performance optimizations both on level of the simulation model and the simulation kernel in section 5. Section 6 provides a brief discussion of profiling a simulation run using internal or external profiler followed by the summary in section 7.

2 Executable Specification of the German Toll System

Toll Collect GmbH is the provider of the German electronic toll for heavy goods vehicles (HGVs). The system automatically collects the toll fees on federal motorways using an on-board-unit (OBU) installed in most of the trucks[1]. Currently there are more than 750 000 OBUs deployed, each determining the toll fees according to an up-to-date map of the chargeable roads using a GNSS receiver coupled to the vehicles' speed and directional data and communicating via GSM with the Toll Collect data center. In total, the HGVs drove $26.6 \cdot 10^9$ km on the chargeable federal motorways in 2012 [8] incurring a total of 4.36 bn € [9]. For the application domain we use an existing simulation model of the German toll system [6, 28], a large-scale autonomous toll system [10].

Following the idea of simulation-driven design we use executable models to analyze and evaluate the behavior of the IT systems of Toll Collect GmbH. The simulation model of the Toll Collect system is used to predict the behavior of the current system as well as effects of changes to the system, especially to maintain the high level of accuracy (with an error rate of less than 1 in 1 000, [12, 36]) needed due to service-level-agreements. Changes to the Toll Collect system occur every day – in the past four years more than 15 major changes (releases) and more than 1 500 medium-sized changes were implemented.

2.1 Modeling of the Toll Collect System

The simulation model of the Toll Collect system consists of three blocks as shown in fig. 1 and an additional model for the user interaction (scenario generator). The model execution is controlled by a discrete event scheduler (in our example either MSArchitect [2] or MLDesigner [24]), responsible to initialize the vehicle fleet and to run the simulation. The vehicle fleet treats each OBU as an individuum with a distinct configuration and internal state. This state changes according to the simulation and the externally pre-calculated (statistically realistic) driving pattern of the

---

[1] An alternate mode of operations is available which offers the ability of manual booking.

OBUs [31, 32] containing their configuration (e.g. hardware and software versions), their power cycles and the toll charging instants.

Starting with the OBU and its driving pattern the model simulates the automatic communication between the OBU and the central systems either to transmit the tolls collected or to update the OBU state, geo and tariff data or software. Due to the arbitrary power cycles of the OBU and various resource restrictions (e.g. limited bandwidth, high latency, intentionally limited number of parallel connections for the central systems) it is common for data transmissions to be interrupted and subsequently recovered by application-level protocols.

The mobile data network includes provider specific transmission properties (e.g. bandwidth and latency) and resource constraints (e.g. actively managed number of simultaneous connections allowed). On the network layer the simulation includes the bandwidths and latencies observed for the various OBU hardware platforms and mobile network operators. The simula-

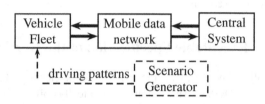

Fig. 1 High-Level simulation model of the Toll Collect system (upper half) and the model for the user interaction (scenario generator, lower half)

tion includes the GPRS connection handling, the authentication handshake and IP address handling but does not include the IP network layer.

The block "Central System" includes the typical systems required to authenticate, receive and validate data transmission (e.g. firewalls, proxy servers, load balancers, database and application servers) each with their individual resource constraints. From a service management perspective the system is a sizeable service value chain spread across several service providers [26, 29, 30, 33].

To achieve realistic simulation results the model tries to include as many details as possible. Accordingly the vehicle fleet should include as many individual OBUs as in the real Toll Collect system (more than 750 000) with statistically realistic driving patterns for several consecutive months. In that way it is possible to simulate long-term behavior (e.g. a software update of the whole fleet) without resorting to scaling. The behavior of each OBU is implemented at a high-level of detail up to including the original source code of the OBU in the handling of internal state transitions.

2.2 Application and Results

The simulation model is used to determine the effects of the systems' configuration, e.g. on the progress of software updates or on the return to normal operations after outages of the central systems. This can be extended to determine the optimal system

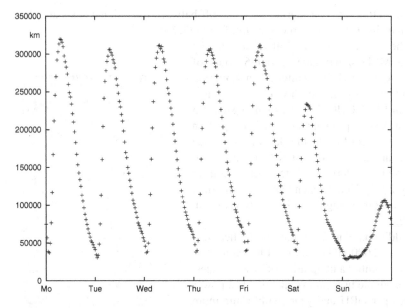

Fig. 2 Weekly driving pattern of a vehicle fleet of 140 000 HGVs. Each point represents the chargeable kilometers driven within a 30 minute period.

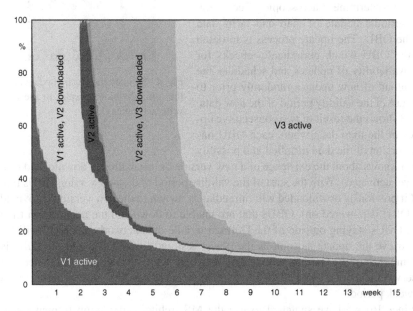

Fig. 3 Two consecutive map data updates of a fleet of 140 000 OBUs over a 15 week period, the update is downloaded and activated only after the start of the validity period

configuration e.g. considering the trade-off between operational costs and the cost of financing (of not yet processed toll fees) [27, 28].

The execution of the Toll Collect model using MLDesigner takes about 6 hours of single-core CPU time (Intel Xeon X5670 at 2.93 GHz) to simulate 16 weeks with a fleet size of 140 000 HGVs (corresponding to a 1:5 scaling). The pre-calculated driving pattern changes according to the day of week (see fig. 2) and is based on a statistical analysis of the driving pattern over a 15 week period (in early 2011). The weekend and Sunday truck ban on German highways is clearly visible in fig. 2.

Additional data from the Toll Collect test fleet (> 2 000 HGVs) is used to parameterize the number and duration of power cycles. The example uses an average of 1112 power cycles per OBU and year (with a minimum duration of one minute per power cycle).

This microscopic simulation model is also used to determine macroscopic effects, e.g. the periodic update of map and tariff data on the OBU. The update process is initiated by the OBU which periodically checks for the availability of updates and schedules the download of new updates randomly prior to the start of the validity period of the new data. Fig. 3 shows the result of two consecutive updates of the map data, where each OBU has one version of the data installed and possibly

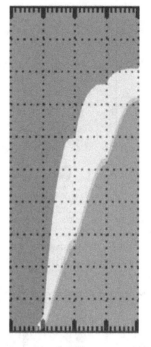

Fig. 4 Simulated software update from the initial version (purple) to the new version (orange) [31]

either knows about the existence of a new version or has it already downloaded (but not yet activated). With the start of the validity period of the new version OBUs that had it previously downloaded will immediately switch to the new version (provided the OBU is powered on). OBUs that are unable to download the new data in time (e.g. OBUs staying outside of the German mobile data network coverage) will try to retrieve the update as soon as the power restored to the OBU and the OBU is within reach of the German mobile data network. Across the whole vehicle fleet we observe that about 10% of all OBUs do not connect to the data center within a given 15 week period.

Since [6] we have switched to use the MSArchitect simulation framework to achieve simulation runs at a 1:1 scale: From the process perspective the simulation model covers business and system processes differing at least 7 orders of magnitude in time: All major technical processes with durations of one second and longer are included in the model aiming to predict the dynamic system behavior of fleet-wide updates (taking weeks to months, fig. 4). In fact, the model includes some processes with higher temporal resolution (down to 50 ms for the connection handling in the DMZ) and is used to simulate all updates occurring over a whole year. Using the Pearson correlation as metric to compare the simulation results with the observed update rates between April 2012 and January 2013 we find the correlation to be above (better than) 0,994 (see tab. 1). The current investigation is to validate the simulation model using additional metrics and a time-scale of one hour [32] (instead of one day).

Even on the application level the user interaction (scenario generator) creates a large number of events to be processed by the simulation logic. On average each OBU will be powered-on for 16% of the time and process tolls for 32 000 km annually ([7], one toll event per 4.2 km on average [12]) spread across some 1 300 power cycles (including three times as many periods of mobile data network). Of course, many more events are created from within the application logic, e.g. to forward tolls to the central systems or to run error recovery protocols in the case of network unavailability.

Table 1 Comparing fleet-wide updates (simulation results vs. data from Apr 2012 to Jan 2013)

	correlation
software	0.99963
geo data	0.99572
tariff data	0.99475

2.3 Simulation Performance

To achieve realistic simulation results we decided against the use of a simplified simulation model (as compared to the real-world system) and aim for a 1:1 scale, i.e. more than 750 000 individual OBUs within the simulation and a realistic behavior on the network layer. Therefore the typical time-scales within the simulation are on the order of 100 ms. However, the business processes of interest have a typical time-scale of one to two months: e.g. map and software updates are intentionally spread over many weeks to be able to reach HGVs that are operating outside of the German mobile network coverage.

As a consequence the simulation performance must allow to simulate at least three consecutive months of a realistic driving pattern with a full-size vehicle fleet. Using the simulation as part of the design process or to validate changes to the systems' configuration necessitates that a typical simulation run delivers results within the business day. Unfortunately, the tools used do not yet allow the automatic distribution of the simulation across several CPUs (or even CPU cores).

The initial implementation of the simulation model with MLDesigner led to various performance bottlenecks due to the large number of OBU objects and scheduled events within the simulation. The extraction of the OBU logic from the simulation model to conventional C++ classes alleviated the performance degradation and the memory usage. Changing the model implementation and switching to the MSArchitect simulation framework we were able to increase the fleet-size to realistic scales. A prerequisite is a detailed understanding of the performance issues present in the simulation model and the tools used for execution. Therefore the remainder of the article focuses on benchmarking of simulation tools or identifying performance hotspots in a given simulation model.

3 Evaluation of Discrete Event Simulation Performance

3.1 Importance of Performance

It is well known [20, 21] that software-intensive systems evolve towards ever increasing complexity – fulfilling more user requirements, interfacing with additional other systems and of course requiring ever more lines of code. Modeling and simulation methodologies and technologies [5] can be applied to design, analyze, evaluate, validate and optimize such systems – far in advance of the actual implementation. Executable models are created as blueprints of the new system and are used as functional ("virtual") prototype. At an early stage of the design process these virtual prototypes give insight into the systems' behavior e.g. regarding the scaling properties, the advantages and disadvantages of the system topology. At any time simulations can be used to explore operational scenarios (especially those exceeding the systems' specification) and the inherent risks (operational and procedural).

In this context simulation performance needs to keep up with the enormous complexity increase of executable models, which in turn follows the complexity increase of systems and processes. In addition, executable models should include a high level of detail. Together with systems including a large number of active components (e.g. users, machine-to-machine networks) this results in a complex simulation model – both from a static and dynamic perspective.

In a business context, the simulation is often part of an optimization process, i.e. the optimal solution is determined by iterative optimization loops. In that case many steps consisting of a complete simulation run (possibly including test batteries) need to be evaluated to determine the optimal solution. Of course this approach is beneficial only if the simulation results are both reliable and available well in advance of traditional software engineering approaches. Hence simulation performance in terms of speed and memory consumption and its benchmarking became a critical aspect in system design.

3.2 Performance Benchmarking

There exist several approaches for benchmarking simulation performance, especially kernel benchmarks and application benchmarks are common [34, 35]. A kernel benchmark consists of several, typically smaller test cases where each test case stresses a single elementary function of the simulation kernel (see fig. 5). Therefore kernel benchmarks are useful to analyze the built-in performance of low-level mechanisms. The results are typically weighted according to their importance for a given application domain – however, the predictive power of kernel benchmarks for real-world application performance is limited. To compare application performance, the benchmark measurements include a number of real-world examples from the application domain. The selected applications should exhibit different characteristics and represent typical challenging workloads.

In the Toll Collect example we started with an existing simulation model using a given simulation tool (MLDesigner). To achieve the necessary performance (as outlined in section 2.3) both kinds of benchmarks were used: A low-level analysis of the simulation kernel allows to identify performance bottle-necks in the existing simulation model and tool. In addition the kernel benchmark is easily

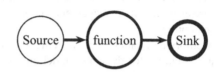

Fig. 5 Basic test model for kernel-level benchmarks used to test elementary functions

adopted toward different simulation tools. Section 4.2 gives a description of the kernel benchmarks used to benchmark a total of five different DES simulation tools, followed by a comparison and discussion of the kernel benchmark test results.

As a consequence of the kernel benchmark results the Toll Collect simulation model was ported to a second simulation tool – requiring considerable effort and expertise (both of the simulation model and tool). Having the same simulation model implemented for two different simulation tools allows for direct comparison and benchmark at the application-level (as shown in section 5.1).

4 Kernel-Level Benchmarks

DES simulations are typically split into the simulation environment and the simulation model. The simulation environment itself is used to create models (using an interactive and graphical user interface), to execute existing simulation models and possibly also to visualize the progress and results of a simulation run. The simulation model itself contains all static model entities, their relationships and methods to handle events during the model execution. Thus the simulation model determines the dynamic properties of a simulation run, e.g. the number of entities present during the model execution and the number of events created.

4.1 Simulation Kernel Benchmark Tests

Since the execution control resides with the simulation kernel, the implementation of event handling (especially the future event list (FEL) and its update mechanism), the data and memory handling (e.g. pass-by-reference vs. pass-by-value, garbage collection) and the use of caches determines the simulation kernel performance. Similar to [13] we include the following elementary factors in our kernel-level benchmarks:

- FEL management: The event scheduling mechanism is the core of any discrete event simulation determining the dynamic behavior of the simulation. At any given time the model entities create new events scheduled to take place in the future and sometimes cancel existing future events as well. The crucial performance factor of a DES simulation kernel is therefore the handling of the future event list. Its management can be more time consuming than the actual data manipulation.
- Memory and data type management: The allocation and maintenance of tokens and memory for dynamic model entities is an important issue. The event handling will inevitably deal with the creation and deletion of a large number of events, events passed between model entities usually need to transport additional (application-level) data between the entities, possibly necessitating the casting between data types (incurring an additional overhead). The efficient storage of the information will directly affect the simulation performance. A *pass-by-value* approach will incur additional overhead (due to the necessity of duplicating the data). A *pass-by-reference* implementation of the FEL management algorithms processing the tokens representing an events should yield better performance – especially if the simulation entities are only referenced from the event tokens. Dealing with memory allocations can be improved by the use of caching mechanisms.
- Pseudo-random number generator performance: A basic requirement of DES simulation execution is the ability to use random numbers to achieve a "non-deterministic" behavior. A typical DES tool includes generators for several different random number distributions. It is critical to be able to use large streams of pseudo random numbers.
- Arithmetic operations: The actual data manipulation is given by arithmetic operations either in an imperative or functional language. This programming language needs to be executed at runtime and can become a performance bottleneck if the chosen programming language does not allow compilation to the underlying CPU architecture.

In addition the ability to generate reports or to export reporting data is a basic requirement for any DES simulation. Creating the reports and the underlying data can incur considerable additional computational expense. However, the reporting requirements are typically driven by the application domain. Therefore we do not include reporting in our kernel-level benchmark.

4.2 Simulation Kernel Performance Tests

We present five different test models for DES simulation kernel benchmarks, addressing all elementary factors presented in section 4.1. These models are applied later to investigate and compare DES kernel performance. The models have been kept simple in order to assure universality regarding different kernels/tools and to avoid possible side effects.

4.2.1   Simulation Scaling

A simulation model with several hierarchy levels (fig. 6) is simulated in a sequence increasing the total number of events, while the size of the future event list remains fixed. This test determines whether the FEL performance is affected by the FEL size. The test used a clock interval of one and the number of events processed increased from $0.1 \cdot 10^6$ to $10\,000 \cdot 10^6$ events.

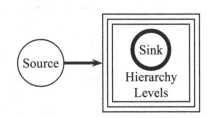

Fig. 6   Test model for simulation scaling

4.2.2   FEL Size Scaling

The second test uses the generic simulation model of fig. 5 with a delay-function. The delay is used to easily configure the (average) number of events waiting in the future events list with minimal variance, while the total number of events processed remains constant. This test examines the overall performance of the FEL algorithm and data management. We used a uniform distribution of events in the FEL list. Of course, the test can be extended toward non-uniform events distributions, in order to check adaptability of the FEL algorithm on different events densities.

For this test we use a clock interval of one, a fixed number of processed events $(300 \cdot 10^6)$ and configure the delay-function to produce a given size of the future events list (between 10^6 to 10^7 events, constant over single experiment).

4.2.3   FEL Adaption

This test extends the future events list size during one test run by changing the parameter of the delay-function dynamically during the simulation execution. The test extends the previous test model by additional single events used to change the parameter of the delay-function (see fig. 7). As a consequence the size of the FEL changes during the test run forcing the simulation kernel to adapt the FEL size (e.g. allocating and deallocating memory) during the simulation run.

The test uses a clock interval of one and a dynamic delay-function parameterized to give a dynamic FEL size of $1\,000 - 10^6 - 100 - 10^7 - 10$ events during the simulation run. In total one test run consists of $200 \cdot 10^6$ processed events.

4.2.4   Memory and Data Type Management

The test creates large data arrays of different sizes and passes the data through the simulation model in sequential or parallel order as depicted in figure 8. When executing the model the memory management of the simulation kernel should recognize the passing of unmodified data and use references to this data. Ideally only one datum should be created and send as reference through the model. As

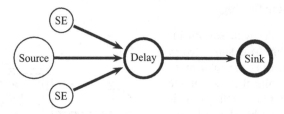

Fig. 7 Test model for FEL adaption with additional single events (SE)

long as the delays are set to zero, no difference between serial and parallel passing should be recognizable. The test uses a fixed number of nodes (delay blocks) either in a parallel or serial configuration and data arrays with $1, 0.5 \ldots 2 \cdot 10^6$ entries.

4.2.5 Random-Number Generator Performance

A large number of random values is generated using different distributions. The model uses a constant function as a reference to measure relative performance of the built-in pseudo-random-number generators. The test computes $20 \cdot 10^6$ random numbers of different random number distributions (normal distribution, Poisson distribution and exponential distribution).

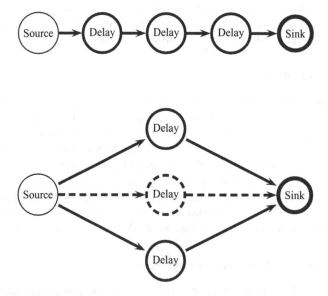

Fig. 8 Test model for memory and data type management for sequential (top) and parallel (bottom) processing

4.3 Evaluation of Simulation Kernel

Each test model is simulated with a set of simulation parameters using different system design tools. Currently more than 80 tools listed for DES [1]. We selected six system design tools for evaluation: Ptolemy II, Omnet++, AnyLogic, MLDesigner, SimEvents and MSArchitect. All tools were run in serial mode (DES, not PDES) on an Intel Core i7 X990 at 3.47 GHz with 24 GiByte RAM using either Windows 7 Enterprise (64 bit) or openSuse 11.4 (32 bit, kernel 2.6.37.6). Performance data was recorded with Perfmon on Windows and sysstat and the Gnome System Monitor on the Linux system.

Fig. 9 gives the results of the Runtime Scaling test. The upper chart shows the event processing performance for different simulation lengths and the bottom chart the private memory consumed during simulation. The tests show that neither the

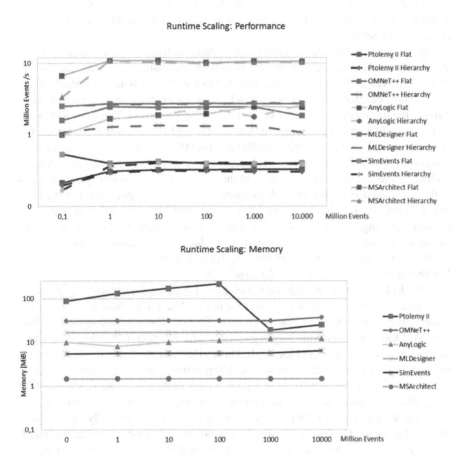

Fig. 9 Runtime performance (top) and memory usage (bottom) scaling of different DES tools

memory consumption nor the event processing performance is affected by increasing the simulation runtime. Looking at the sensitivity of running the tests with additional hierarchy levels we find that only OMNeT++ is sensitive to the additional hierarchy levels. However, from the test results it is already obvious that the different tools vary in event processing performance by an order of magnitude: MSArchitect provides the highest speed. MLDesigner, AnyLogic and OMNeT++ provide 25% of the speed (compared to [6] the MSArchitect performance improved by more than 30%). Ptolemy II is twenty times slower. Looking at the memory usage during the simulation the difference between the tools is again more than an order of magnitude – the slowest tool using the most memory and the fastest tool using the least. Two of the tools (Ptolemy II and AnyLogic) are based on the Java programming language, where explicit memory deallocation is not possible. Apparently the Ptolemy II test run triggers the JVM garbage collection during the simulation run and is able to free 90% of its memory. As a result Ptolemy II memory consumption is then comparable to the next three simulation tools. The AnyLogic test run starts already with a much lower memory consumption than Ptolemy II and no effect of JVM garbage collection is visible.

Fig. 10 gives the results of the FEL Size Scaling test. As expected, a systematic performance decrease can be observed with increasing FEL size, due to the increasing overhead for FEL management. Most of the tools tested initially start with relative constant performance (on a log-log scale). With increasing FEL size three of the five tools develop drastic performance degradation. This coincides with a rapid grow of memory consumption with increasing FEL size. We propose that the performance reduction is correlated with increased FEL memory usage due to a performance penalty of calendar queue based schedulers for large queue sizes. Again, Ptolemy II has the lowest performance in this test. OMNeT++ is nearly not affected in the considered FEL size interval. In absolute numbers, MSArchitect has the best test performance and the lowest memory usage until FEL size 10^6. Subsequently the memory usage of MLDesigner is lower since MSArchitect runs in 64 bit mode which in fact means a higher memory demand due to larger address ranges. But in our benchmark MLDesigner stops working for FEL sizes above $15 \cdot 10^7$. We tested MSArchitect successfully with a FEL size of 10^8.

In the FEL Adaption test the simulation kernel is subjected to a varying demand to its FEL. Beyond the runtime needed for the test the main result is the memory consumption during the test run as given in fig. 11. The simulation took 2 276s with MLDesigner, 673s with AnyLogic, 192s with MSArchitect, 395s with OMNeT++ and over 2 hours with Ptolemy II. During that time the dynamically changing memory usage varies widely between the different tools. Most tools tend to allocate memory in chunks visible as steps in fig. 11. Again, Ptolemy II is the slowest tool in comparison and also requires more memory than any other tool in the benchmark. The memory usage of OMNeT++ indicates the ability to dynamically free already allocated memory. However, this simulation tool also allocates considerably more memory than any of the other tools for a brief period of time during the test run.

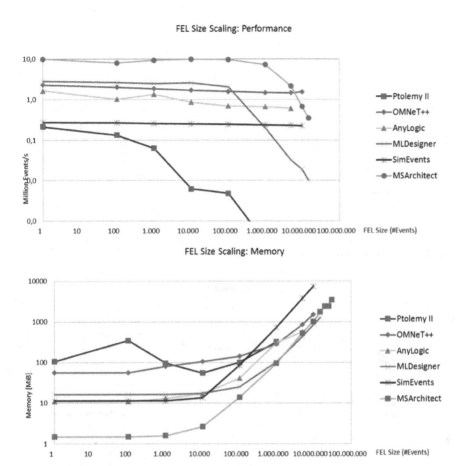

Fig. 10 Future event list size scaling test results for runtime performance (top) and memory usage (bottom) scaling of different DES tools

The Data Type Management test passes large arrays of data through the simulation running either in a parallel or serial configuration. The memory consumption during the test run is shown in fig. 12. Most tools handle serial and parallel passing of token data in a different way, which can be recognized by the gap in memory consumption between both serial and parallel versions. Ptolemy II and MSArchitect do not show a difference between the parallel and serial version, only references are passes when only delays are used. However, Ptolemy II requires more memory and shows a different behavior according to the memory allocation: a large portion of the memory is allocated at initialization time with standard modeling elements.

The last test is the Random Number Generation test. As depicted in fig. 13 the performance does not depend on the type of the generated distribution, since there are only minor differences to the generation of constant numbers. Again, Ptolemy II is an order of magnitude slower than OMNeT++, AnyLogic and MLDesigner in this

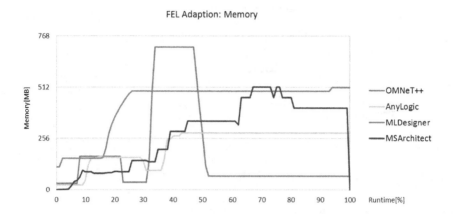

Fig. 11 Test results for future events list adaption

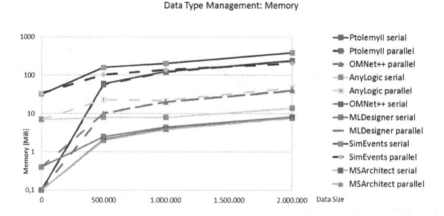

Fig. 12 Test results for the memory consumption during Data Type Management test case

test. MSArchitect gives the highest performance compared to the other tools. It is not clear whether the pseudo-random number generator (PRNG) algorithm differs between the five simulation tools or if the PRNG performance is adversely affected by event management overhead. Since this test relies on the correct implementation of the PRNG, i. e. we do not check the statistical quality of the random numbers generated, the test results might not be fair if one of the tools were to use low-quality but high-speed generators.

It can be concluded, that Ptolemy II is inferior in all simulation kernel benchmarks performed. MLDesigner is equal to or better than AnyLogic in all categories but FEL adoption. Due to the utilization of the JVM, AnyLogic requires more memory in equivalent models and therefore scales worse with increasing FEL size. Both,

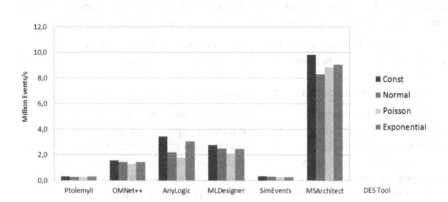

Fig. 13 Test results for random number generator scaling

MSArchitect as well as OMNeT++ show the best performance in some categories. MSArchitect is the fastest simulation kernel in most categories and requires least memory for data handling.

5 Enhancing Simulation Performance of the Toll System Model

While section 4 focused on the performance evaluation of DES kernels we now focus on the application-level performance, i.e. how to specify *efficient* executable DES models.

5.1 Evaluation of Model Architecture

A simulation model can be thought of as a (simplified) copy of an existing or imaginary system, created for a certain purpose. The model and the process of creating the model are a key to learning and communicating about the system itself. This implies that the right level of abstraction needs to be found so as to include only the system behavior relevant to the models' purpose. Bearing this in mind the most important rule in designing efficient models can be derived: *The level of detail always follows the model purpose.* For instance it makes a huge difference for choosing the appropriate level of detail when designing a data transmission model compared to modeling a rather abstract business process. Of course, any useful model must be connected in some way to the reality. A second point directly connected to that rule is to focus on measurable system behavior. Otherwise the model could become worthless when using it for analyzes and optimization.

Designing the simulation model directly affects the runtime properties (e.g. performance and memory consumption). From a technical point of view the performance of a model can be improved considerably by addressing several issues:

- the number of simulation entities present during the simulation run,
- the data transport between model components to reduce the number of DES events and
- the execution time and memory consumption of the model components.

Hence whenever possible, highly interacting model components should be merged together to avoid the time consuming data exchange via the simulation kernel. In addition data should always be transmitted in form of references/pointers (*pass-by-reference*). References are values that enable indirect access to a particular datum, indistinct from the data itself. They are used to efficiently pass large or mutable data. In that way the time-consuming and unnecessary copying of data is avoided.

The simulation kernel benchmarks in the previous section identified the future events list as a key factor in the kernel performance. The tests were designed to continuously create new events leading to different FEL sizes. Creating, scheduling and passing events is certainly the key feature of DES simulation kernels. However, a simulation model sometimes needs to be able to cancel scheduled future events before they are executed. Many simulation tools lack a good implementation for canceling events from the FEL, possibly needing to traverse the whole FEL in the search of the canceled event and possibly triggering memory reorder after removing the canceled event from the FEL. Obviously, simulation runs with a large FEL are more affected.

The performance of simulation models can be improved by transferring part of the simulation model to existing standardized model components or even extending the simulation tools' existing catalog of standard components.

5.2   Performance Enhancement to Our Solution

As the project of modeling the German toll system was launched our team had no clear picture of the coming performance issues: Existing simulation models of HGV tolling systems both at Toll Collect and in the literature were limited to a few thousand simulated HGVs [16, 22] and reaching 500 000 HGVs over a 4 week period [23]. Our model aimed to include a more detailed behavior and a vehicle fleet almost two orders of magnitude larger (comparable but still larger than simulations of metropolitan car traffic, e.g. [14] using 200 000 drivers with a shorter simulated time frame).

After putting together and validating the basic DES model in MLDesigner, including the dynamic behavior of the vehicle fleet, mobile providers and central system we tried to scale up to the real world situation. This meant to run simulations of vehicle sizes of up to 750 000 HGVs over a simulated time period of at least 3 months. The disappointing simulation performance results are shown in the second column of table 2. The desired scenario took about 49 million seconds, over 6 times *slower than reality*, an unacceptable result.

By transferring the model from the system design tool MLDesigner to MSArchitect the simulation performance could be increased dramatically. On the one hand the throughput of events is about 3.5 to 4 times higher in MSArchitect (as confirmed by the kernel benchmarks in section 4). On the other hand we recognized huge performance issues in the management of complex data structures in MLDesigner. We analyzed the differences by comparing the simulation performance of using MLDesigner data structures versus using pointers to external C++ classes for data transport (the default in MSArchitect). In total the transfer of our simulation model from MLDesigner to MSArchitect brought a 120-fold speed increase.

Table 2 Simulation performance for the Toll Collect example with a simulated time period of three months

| | Runtime [s] | |
fleet size	MLDesigner	MSArchitect
70 000	0.25 M	900
700 000	49.00 M	8 700

On top we redesigned our model architecture. First of all removing all "cancel event" operations from the model – being rather expensive operations in both simulation tools. The canceling of events was replaced by introducing an additional boolean tag to store whether the next receiving event is ignored or not. By doing so the overall amount of events in the FEL is increased and more memory is needed but time consuming cancel operations can be avoided.

Next we removed several retry processes between vehicle fleet and mobile data network providers and merged heavily interacting model components to minimize data transport across the simulation kernel. In addition we switched to the data structure mechanism of MSArchitect which automatically uses references when sending or receiving unchanged data tokens.

In total a further significant performance increase could be achieved. The rightmost column of table 2 shows the results of two different scenarios executed with MSArchitect. Simulating the scenario stated above (750 000 HGVs over a three months period) took about 8 700 seconds. Thus the simulation speed could be increased by a factor of 5 630 compared to the initial runs using MLDesigner. It is noteworthy that the model used with MSArchitect also includes additional additional functionality of the German toll system.

6 Profiling of the Simulation Model

To evaluate the application-level simulation performance of our model of the German toll system, we use both the kernel logging capabilities of MSArchitect and an external profiling application (Intel VTune). Kernel logging allows to count the number of calls of atomic models as well as the total number of samples (corresponding to a processor cycle). The external profiler allows measuring the space complexity (memory), the time complexity (duration, CPU time) and the usage of particular instructions of a target program by collecting information on their execution. The most common use of a profiler is to help the user evaluate alternative implementations for program optimization. Based on their data granularity, on how profilers collect information, they are classified into event based or statistical profilers [15]. We've selected the statistical profiler Intel VTune Amplifier XE and connected it to the generated C++ runtime representation of our model. As test environment, an Intel Core i7 K875 at 2.93 GHz with 8 GiByte RAM and Windows 7 Professional (64 bit) installed has been used. To profile the simulation model we take the simulation scenario used to verify the simulation model against real-world data (Apr 2012 to Jan 2013).

6.1 Profiling with MSArchitect

In a first step we apply the kernel logging capabilities of MSArchitect resulting in a file with profiling information at the end of the simulation run. Tab. 3 shows an excerpt of the file, containing all atomic blocks relevant to analysis (15 out of 65). Since during simulation all composite blocks are resolved to directly communicating atomic blocks, the table only contains atomic blocks of the simulation model. For each atomic block the table shows the number of calls, the accumulated count of samples, the time required in relation to other atomic blocks and the samples needed for one call.

First of all, the atomic block AccessSessionStateSwitch is striking, since it consumes a large amount of time due to the high number of calls. The block is responsible for switching OBU data structures in response to its state to one of the output ports. As the block switches between 34 states, 539 samples per call are acceptable. Nevertheless the number of calls could be reduced for performance

Table 3 MSArchitect kernel performance logging results

Atomic Block	Calls [M]	Samples [G]	Time [%]	Samples per Call
AccessSessionStateSwitch	19 980	10 760	10,89	539
ExternDStxt	0,0007	9 565	9,68	13 665 M
StaHandling	482	6 503	6,58	13 483
EinzelbuchungsHandling	4 660	6 442	6,52	1 382
IpAutomat	7 323	5 749	5,82	785
Delay (Standard)	8 874	5 522	5,59	622
CheckComponentState	7 363	3 859	3,91	524
NetzverlustHandling	3 020	3 479	3,52	1 152
AccessSessionStateWrite	5 841	3 215	3,26	551
MfbSwitch	5 525	3 196	3,24	579
Nutzdaten	3 563	2 694	2,73	756
TcmessageCopy	2 030	2 010	2,03	990
TcpAutomat	1 291	1 533	1,55	1 187
TimedAllocate	2 570	1 484	1,50	578
SimOutObuVersions	0,017	1 509	1,53	89 M

improvement by changing the model architecture – especially once the model is ported to the parallel DES core, it is an obvious block for introducing parallelism.

The next conspicuous atomic block is ExternDStxt, reading the pre-generated files provided by the scenario generator model as ASCII file. The block consumes 13 665 M samples/call and is rarely executed (700 times, i.e. twice per simulated day) resulting in a time consumption of 9,681% of the time. In order to reduce the load, scenarios should be computed on the fly. The atomic block StaHandling is responsible for generating and controlling status requests, which may result in update processes. The block consumes 6,581% of simulation time. We see potential for improvements in changing the implementation (e.g. conversion of formulas to save operations, replacing divisions by multiplications with reciprocal and using of compare functions from standard libraries).

With 4 660 M calls EinzelbuchungsHandling is a frequently executed atomic block. After analyzing the implementation we find 1 382 samples/call acceptable. The block depends on the random number generator and would benefit from faster random number generation algorithms. The atomic block SimOutObu Versions cyclically writes the software, region and tariff version of all OBUs to an output file. In our scenario we simulate 50 weeks and write data every 30 minutes, resulting in 16801 calls. 89 M samples/call seems to be quite costly and offers room for improvement.

In summary the simulation of the scenario took 98 811 263 M calls. Of these, the model components consumed 84,51% and the simulation kernel (logical processor) 15,49%.

6.2 Profiling with Intel VTune

In the second step we apply the profiling application Intel VTune [18]. The external profiler catches the activities of both the simulation kernel and the simulation model (denoted as "K" or "M" in tab. 4).

An excerpt of the results is shown in tab. 4. For each function the CPU time in percent, the amount of needed instructions (instructions retired), the estimated instruction call count, the instructions per call on average and the last level cache miss rate (0,01 means one out of one hundred accesses takes place in memory) is shown.

Most of the CPU time is consumed by kernel functions responsible for data transport. These functions are grouped by component (resp. namespace msa.sim. core, denoted as "K" in the first column of tab. 4). In total these functions consume 61,1% of the CPU time. Conspicuous is the relative high last level cache miss rate of function EventManager.enqueueEvent with 3,2% and the number of instructions needed per call LogicalProcessor.mainLoopFast with 2 379. However, the number of calls depends on the dispatch of data within atomic model components, which are grouped in form of user libraries. In our model we have two user libraries: GPRSSimulation (GPRSSimulation.Components.Atomics, denoted as "M" in the first column of tab. 4) and Standard (msa.Standard.

Table 4 VTune profiling results for simulation kernel (K) and model (M) ordered by CPU time. Shown are the CPU instructions retired (IR), estimated call count (eCC), instructions per call (IPC) and last level cache miss rate (MR).

Function	Time [%]	IR [G]	eCC [M]	IPC	MR [%]
K Port.send	9,0	44	689	65	0,4
K EventManager.enqueueEvent	7,7	21	92	237	3,2
K LogicalProcessor.mainLoopFast	7,0	17	7	2 379	0,3
K EventManager.dequeueEvent	6,3	104	2 517	41	1,1
K big._mul<unsigned int>	5,0	103	2 611	40	0,3
M StaHandling.Dice	4,8	12	11	1 097	0,1
K EventManager.scheduleEvent	3,5	53	1 286	42	0,2
K Any.extractToken	3,0	70	1 805	39	1,7
K Pin.popFrontToken	2,8	49	1 234	40	0,2
K EventManager.bucketOf	2,7	17	327	55	0,0
K Any.operator=	2,5	54	1 403	39	0,2
K Any.create	2,3	64	1 689	38	0,4
K random.tr1.UniformRng.getNextV	2,2	39	961	41	0,2
K Any.doClear	2,1	22	497	45	0,2
K Tokenizer.nextToken	1,8	22	606	36	0,3
K TemplatePort<Tcmessage>.receiveToken	1,7	29	726	40	0,3
K TemplateTypeInfo<EventData>.createToken	1,6	84	2 326	36	2,1
M AccessSessionStateSwitch.run	1,5	13	287	48	0,2
M EinzelbuchungsHandling.run	1,4	5	66	87	7,2
M IpAutomat.run	1,4	7	103	70	3,4
K Pin.popFront	1,3	6	89	74	0,3

Control). The latter is a support library included in MSArchitect. Combined they are responsible for 20,1% of CPU time consumption. Performance critical and starting point for improvement is the function StaHandling.Dice with 1 097 instructions per call and a CPU time consumption of 4,80%.

Both, kernel logging and profiling showed that most of the resources are utilized by functions responsible for data input/output (data mining) and functions responsible for transmission and processing of tolling information. By doing the analysis we located multiple components with potential for optimization, e.g. Access SessionStateSwitch and StaHandling. Furthermore we came to the conclusion to generate scenarios on the fly since the reading of pre-generated scenario files is as time consuming. Relating the resource utilization of model components to real-word applications we could recognize a weak correlation. Model components like STAHandling, EinzelbuchungsHandling and IPAutomat are abstractions of important real word system applications and crucial to performance in both worlds.

7 Summary and Outlook

Extending [6] we have shown how to analyze the performance of DES simulations: Generic benchmark test-cases allow a simple and direct comparison of different simulation tools. Not surprisingly the tools differ vastly as to their time and memory consumption. However, the benchmark results cannot be transferred to the application domain: The workload generated by a given simulation model determines in large part its performance. Taking an existing simulation model of a large-scale technical system we performed an in-depth performance analysis for one simulation tool using both the performance analysis methods provided by the simulation kernel and an external profiler with access to the CPU hardware profiling support.

Both profilers immediately identify the same bottleneck: Reading the ASCII-formatted pre-calculated driving patterns from disk. Further analysis showed that calculating the driving patterns is less time-consuming than storing them on disk. Consequently the simulation model is now integrated with the scenario generator. This in turn will allow implementing an optimization algorithm to fit the driving patterns to the observed system behavior – a feature that we expect to drastically improve the accuracy of the simulation results for the short-term behavior [32].

The hardware profiler catches both the application-level methods as well as the atomics provided by the simulation kernel (with or without access to its source code). Taking the workload generated by this application we can start to tune the behavior of the atomics to improve the overall performance. Looking e.g. at the cache miss rate we find some simulation kernel routines and several application-level methods with a considerable probability of needing access to the main memory. We take this as starting point for future improvements.

MSArchitect, the simulation kernel used in the application benchmark, is currently extended to allow the automatic model reduction and (semi-) automatic parallelization of simulation runs. The single-core benchmark performed here will be the baseline to measure the improvements against.

References

[1] Albrecht, M.C.: Introduction to discrete event simulation (2010), http://www.albrechts.com/mike/DES/Introduction (accessed April 10, 2012)

[2] Andato GmbH & Co. KG: MSArchitect, http://www.andato.com/ (accessed May 12, 2013)

[3] Banks, J., Nelson, B.: Discrete-Event System Simulation. Prentice Hall (2010)

[4] Baumann, T.: Automatisierung der frühen Entwurfsphasen verteilter Systeme. Südwestdeutscher Verlag für Hochschulschriften, Saarbrücken (2009)

[5] Baumann, T.: Simulation-driven design of distributed systems. SAE Technical Paper (2011-01-0458) (2011), doi:10.4271/2011-01-0458

[6] Baumann, T., Pfitzinger, B., Jestädt, T.: Simulation driven design of the German toll system – evaluation and enhancement of simulation performance. In: 2012 Federated Conference on Computer Science and Information Systems (FedCSIS), pp. 901–909. IEEE (2012)

[7] Bundesamt für Güterverkehr: Maut-Jahresstatistik 2011/2010 (2012),
http://www.bag.bund.de/SharedDocs/Downloads/DE/Statistik/
Lkw-Maut/Jahrestab_11_10.pdf?_blob=publicationFile
(accessed March 10, 2012)

[8] Bundesamt für Güterverkehr: Maut-Jahresstatistik 2012/2011 (2013),
http://www.bag.bund.de/SharedDocs/Downloads/DE/Statistik/
Lkw-Maut/Jahrestab_12_11.pdf?_blob=publicationFile
(accessed June 10, 2013)

[9] Bundesministerium der Finanzen: Sollbericht 2013. Monatsbericht des BMF 2, 6–57
(2013), http://www.bundesfinanzministerium.de/Content/DE/
Monatsberichte/2013/02/Downloads/monatsbericht_2013_02_
deutsch.pdf?_blob=publicationFile&v=4

[10] CEN: ISO/TS 17575-1:2010 electronic fee collection - application interface definition
for autonomous systems - part 1: Charging (2010)

[11] Coulouris, G.F., Dollimore, J., Kindberg, T., Blair, G.: Distributed Systems: Concepts
and Design. Addison-Wesley (2011)

[12] Dettmar, M., Rottinger, F., Jestädt, T.: Achieving excellence in GNSS based tolling
using the example of the german HGV tolling system. In: Proceedings of the 9th ITS
Europe Congress (2013)

[13] Fishman, G.S.: Discrete-Event Simulation: Modeling, Programming and Analysis.
Springer, Berlin (2001)

[14] Flötteröd, G.: Traffic state estimation with multi-agent simulations. Ph.D. thesis, TU
Berlin (2008)

[15] Graham, S.L., Kessler, P.B., Mckusick, M.K.: Gprof: A call graph execution profiler.
ACM Sigplan Notices 17(6), 120–126 (1982)

[16] Hericko, M., Hericko, M., Zivkovic, A.: An evaluation of different functional solutions
for satellite-based tolling in europe. In: Hawaii International Conference on System
Sciences, pp. 1–10 (2011), doi:10.1109/HICSS.2011.51

[17] Institute, E.S.: European user survey analysis. Report USV EUR 2.1 (1996)

[18] Intel: Intel VTune Amplifier,
http://software.intel.com/en-us/intel-vtune-amplifier-xe
(accessed May 12, 2013)

[19] Lee, E.A., Messerschmitt, D.G.: Static scheduling of synchronous data flow programs
for digital signal processing. IEEE Transactions on Computers 100(1), 24–35 (1987)

[20] Lehman, M.: Programs, life cycles, and laws of software evolution. Proceedings of the
IEEE 68(9), 1060–1076 (1980), doi:10.1109/PROC.1980.11805

[21] Lehman, M.M.: The role and impact of assumptions in software development, maintenance and evolution. In: IEEE International Workshop on Software Evolvability, pp.
3–14 (2005), doi:10.1109/IWSE.2005.14

[22] Lunde, K., Kieble, L.: Simulating communication within a satellite-based automated
toll collection system. In: Proceedings of the 55th International Scientific Colloquium,
pp. 318–323 (2010)

[23] Lunde, K., Kieble, L., Funk, M.A.: Prediction strategies in a service level granting
prefetching cache for version-controlled gis data. ISAST Transactions on Computers
and Intelligent Systems 2(2), 46–51 (2010)

[24] MLDesign Technologies, Inc.: MLDesigner (2012),
http://www.mldesigner.com/ (accessed April 10, 2012)

[25] Mosekilde, E.: Topics in Nonlinear Dynamics: Applications to Physics, Biology and
Economic Systems. World Scientific Pub. Co. Inc., Singapore (1996)

[26] Opitz, F., Pfitzinger, B., Jestädt, T.: Service levels of a cost center organization. In: Alt, R., Fähnrich, K.P., Franczyk, B. (eds.) Practitioner Track International Symposium on Services Science (ISSS 2009), vol. 16, pp. 81–86 (2009)

[27] Pfitzinger, B., Baumann, T., Jestädt, T.: Analysis and evaluation of the german toll system using a holistic executable specification. In: 45th Hawaii International Conference on System Sciences (HICSS), pp. 5632–5638 (2012), doi:10.1109/HICSS.2012.111

[28] Pfitzinger, B., Baumann, T., Jestädt, T.: Network resource usage of the german toll system: Lessons from a realistic simulation model. In: 46th Hawaii International Conference on System Sciences (HICSS), pp. 5115–5122. IEEE (2013), doi:10.1109/HICSS.2013.415

[29] Pfitzinger, B., Bley, H., Jestädt, T.: Service catalogue and service sourcing. In: Abramowicz, W., Alt, R., Fähnrich, K.P., Franczyk, B., Maciaszek, L.A. (eds.) Informatik 2010, Business Process and Service Science–Proceedings of ISSS and BPSC, vol. 177, pp. 55–62 (2010)

[30] Pfitzinger, B., Gründer, T., Jestädt, T.: Sourcing decisions and IT service management. In: Alt, R., Fähnrich, K.P., Franczyk, B. (eds.) Practitioner Track International Symposium on Services Science (ISSS 2009), vol. 16, pp. 71–80 (2009)

[31] Pfitzinger, B., Jestädt, T.: Exploring the HGV fleet behavior: Notes from the German toll system. In: Proceedings of the 9th ITS Europe Congress (2013)

[32] Pfitzinger, B., Jestädt, T., Baumann, T.: Simulating the German toll system: Choosing 'good enough' driving patterns. In: für Verkehrstechnik, L. (ed.) Proceedings of the mobil.TUM 2013 – International Conference on Mobility and Transport. Technische Universität, München (2013)

[33] Pfitzinger, B., Jestädt, T., Helmers, W., Kosterski, S.: Best practices im contract lifecycle. In: Auerbach, M., Oecking, C., Jahnke, R., Strecker, F., Weber, M. (eds.) Best Practices im Outsourcing, pp. 185–200. Bitkom (2010)

[34] Ronngren, R., Barriga, L., Ayani, R.: An incremental benchmark suite for performance tuning of parallel discrete event simulation. In: Proceedings of the Twenty-Ninth Hawaii International Conference on System Sciences, vol. 1, pp. 373–382 (1996), doi:10.1109/HICSS.1996.495484

[35] Tewoldeberhan, T., Verbraeck, A., Valentin, E., Bardonnet, G.: An evaluation and selection methodology for discrete-event simulation software. In: Proceedings of the Winter Simulation Conference, 2002, vol. 1, pp. 67–75 (2002), doi:10.1109/WSC.2002.1172870

[36] Toll Collect GmbH: Truck toll system proven effective (2012), http://www.toll-collect.de/en/company/news/truck-toll-system-proven-effective.html (accessed March 11, 2012)

Geoportal as Interface for Data Warehouse and Business Intelligence Information System

Almir Karabegovic and Mirza Ponjavic

Abstract. There is increasing interest of organization for advanced presentation and data analysis for public users. This paper shows how to integrate data from enterprise data warehouse with spatial data warehouse, publish them together to online interactive map, and enable public users to perform analysis in simple web interface. As case study is used Business Intelligence System for Investors, where data comes from different sources, different levels, structured and unstructured. This approach has three phases: creating spatial data warehouse, implementing ETL (extract, transform and load) procedure for data from different sources (spatial and non-spatial) and, finally, designing interface for performing data analysis. The fact, that this is a public site, where users are not known in advanced and not trained, calls for importance of usability design and self-evident interface. Investors are not willing to invest any time in learning the basics of a system. Geographic information providers need geoportals to enable access to spatial data and services via the Internet; and it is a first step in creating Spatial Data Infrastructure (SDI).

1 Introduction

Business users constantly search for new and better ways for improving data warehousing capabilities. Many of them already use existing capabilities to strengthen analytics and business intelligence (BI). They have covered dimensions who, what, when, and why, but, there is only rare answer for where.

The first law of geography according to Waldo Tobler is "Everything is related to everything else, but near things are more related than distant things" [1]. But standard data warehouse cannot answer on the following questions that arise as a result of just this law. How far workers would travel from their house to job? How to choose the best location for a new dam? Optimize delivery route to meet

Almir Karabegovic
University of Sarajevo, Faculty of Electrical Engineering,
Department for Computer Science and Informatics, Bosnia and Herzegovina

Mirza Ponjavic
University of Sarajevo, Faculty of Civil Engineering,
Department of Geodesy, Bosnia and Herzegovina

M. Mach-Król and T. Pełech-Pilichowski (eds.), *Advances in Business ICT*,   27
Advances in Intelligent Systems and Computing 257,
DOI: 10.1007/978-3-319-03677-9_2, © Springer International Publishing Switzerland 2014

changing customer demands and conditions? See which parcels and building are in potential flooding areas?

To answer those and other crucial business questions, we need access to true location intelligence – the kind of intelligence geospatial analysis can deliver [2] [3]. Location Intelligence (LI) has the capacity to organize and understand complex phenomena, through the use of geographic relationships, which are inherent in all information. By combining geographic- and location-related data with other business data, organizations can gain critical insights, make better decisions and optimize important processes and applications. Location Intelligence offers opportunity that organizations streamline their business processes and customer relationships to improve performance and results.

The growing availability of geospatial data and the demand for better analytic insight have helped to move location analysis from limited departmental implementations into enterprise-wide environments; from the hands of geographic information system (GIS) experts to IT organizations for deployment across businesses. Most of data in enterprise data warehouses (EDW) have a location reference. It makes possible that every business can enhance its business analytics with location intelligence as shown in Fig. 1.

The fact is that many of those companies cannot have a complete view of their business because their data are not integrated. Instead, their location data are stored in multiple departmental data marts (silos) that drive up costs, cause redundant data, and most important, does not utilize the richness of their data warehouse [4].

2 Use Case: Investment Promotion

2.1 Business Problem Background

Foreign Investment Promotion Agency (FIPA) of Bosnia and Herzegovina (BiH) is a state agency established with the mission to attract and maximize the flow of foreign direct investment into Bosnia and Herzegovina, and encourage existing foreign investors to further expand and develop their businesses in BiH, as well as facilitate the interaction between public and private sectors. It has an active role in policy advocacy in order to contribute to continually improving environment for business investments and economic development, and to promote a positive image of Bosnia and Herzegovina as a country that is attractive to foreign investors.

The project aims attracting and retaining cross-border investments and to provide better access to available land- related information. International best practice and academic research clearly suggest that easy access to land-related information is a key issue for domestic businesses and international investors. Better access to land-related information is clearly associated with greater government effectiveness and better quality of public services and will ultimately increase levels of

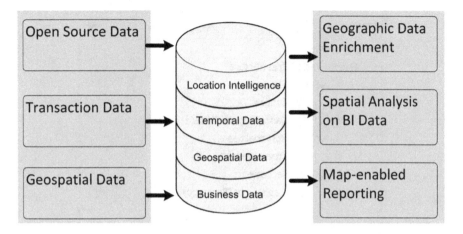

Fig. 1. Location intelligence as Business Intelligence with geospatial capabilities

investment. The World Bank Group's and the European Bank for Reconstruction and Development's Business Environment Survey (BEEPS) database identifies access to land as one of the major concerns for businesses. The European Union's research shows that making various forms of key land and property related information and information on practices, procedures and of relevant laws and regulations available and easy to access through European Union Land Information System (EULIS) is associated with larger average of business and investment opportunities ultimately leading to an improved business environment and investment climate. It also encourages a spread of best practices in presenting land-related data to businesses, establishing basis for comparing performance of localities and stimulates competition among localities in preparing land-related data in a digitalized form and making them available to the public users.

In Bosnia and Herzegovina, various projects have been implemented related to land registration and administration, and land construction resulting in digitalized form of land-related data in different institutions and levels of government. However, this land-related information is scattered in different ministries and agencies at different levels of government, and this makes it difficult for either public or private sector users to obtain land-related information easily and at an appropriate cost and timeframe.

The goal is therefore to build a platform using advanced GIS and web-based database technologies that would make key land-related information needed for investors available and easy to access. The resulting interactive map is intended to become a comprehensive source of land related information relevant for businesses and investors that is easy to view on-line and that can serve a wide range of public and private sector end users too.

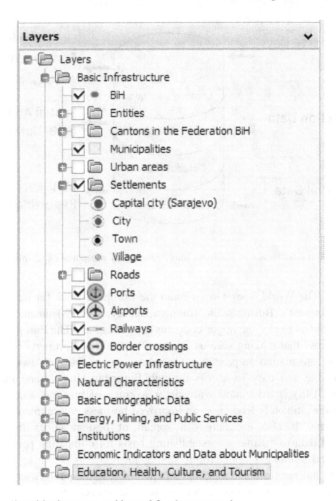

Fig. 2. Layer list with pictogram and legend for the geoportal

Initially, the following layers and data are available within the interactive map, as shown in Fig. 2: basic infrastructure (administrative boundaries, cities, roads, ports, airports, rail, and border crossings); detailed infrastructures (electricity, telecommunication, gas and water supply); natural resources/ environment (climatic zones, land use/ land cover, precipitation, soils, forest canopy coverage, elevation/ digital elevation model, water bodies); basic demographic data per municipality (population disaggregated by gender, age, education, employment, labor force availability); economic data (business entities with addresses, gross domestic product, gross investment, industry data for power plants, mining, manufacturing); special economic zones (business and industrial zones, technological parks and incubators, localities available for investment projects); institutions (business registration courts, customs offices, academic institutions, objects of cultural and historical importance, top touristic locations); and other available data, images or links relevant for businesses and investors.

The task was to analyze and design the new system which will answer their goal. It is recognized that main obstacle was missing of any infrastructure for collaboration and sharing data between different agencies that produce spatial data. Searching for the best practices it is found that existing technologies convergence like web portals, data warehousing and location intelligence could offer to develop the new concept.

3 Geoportals

Geoportal is a web service platform for advanced application development, viewing, and editing of geospatial and business information in a service-oriented architecture. In this use case, solution was based on the Oracle Spatial database and Oracle Middleware Map Viewer web client platform.

This kind of portals can dramatically expand the availability of location based data to non-expert users for re-view, editing, and analysis. It enables fast and efficient creation and configuration of tailor-made, intuitive geodata applications for broad bases of users who require geospatial information integrated with business intelligence [5] [6]. As a result, task oriented, intuitive applications are now available to end users via internet without any software installation on client computers.

From other side, next-generation BI capabilities enable IT and business professionals to effectively leverage spatial analytics, improve system performance, and enhance management of complex BI environments. It means that technologies: Spatial Mapping capabilities with Business Intelligence analytic capabilities, work together to make better business decisions with automated, integrated location intelligence. This means that spatial information needs data appliances that can handle the volume and process it with BI technology and customers are demanding location-based data analytics.

Based on previous statements, it is possible to conclude that geoportal is a type of web portal used to find and access geographic information and associated geographic services (display, editing, analysis, etc.) via the Internet. Geoportals are important for effective use of GIS and a key element of Spatial Data Infrastructure (SDI).

Geographic information providers, including government agencies and commercial sources, use geoportals to publish descriptions (geospatial metadata) of their geographic information. Geographic information consumers, professional or casual, use geoportals to search and access the information they need. Thus geo-portals serve an increasingly important role in the sharing of geographic information and can avoid duplicated efforts, inconsistencies, delays, confusion, and wasted resources.

Recently, there has been a proliferation of geoportals for sharing of geographic information based on region or theme. Examples include the INSPIRE, or

Infra-structure for Spatial Information in the European Community geoportal, and UNSDI, the United Nations Spatial Data Infrastructure.

Modern web-based geoportals include direct access to raw data in multiple formats, complete metadata, online visualization tools so users can create maps with data in the portal, automated provenance linkages across users, datasets and created maps, commenting mechanisms to discuss data quality and interpretation, and sharing or exporting created maps in various formats. This empowers BI solution with complementary technologies including spatial ETL, data visualization, and geographic information systems. There are many use case examples of fields that can use advantages of such solutions like Government Healthcare (Disease Outbreak Tracking), Retail (Trade Area Analysis) or Marketing (Location-based Marketing Effectiveness) [7] [8].

3.1   Used Technology

In this project, it is chosen Oracle technology, because its database performance, enterprise data warehousing and analytics, which makes it easier to move location data into EDW for analysis.

This is a robust solution combining database native geospatial capabilities, geospatial services, and integrated analytics. It stores geospatial data directly in database environment so it is managed with business data. This provides consistency and integration and allows analytical tools to access geospatial data along with the rich data in data warehouse.

Geospatial technology is an in-database capability, and there are no extra costs or data marts required to use the capability. It integrates geospatial data and functions in data warehouse, which add benefits of in-database processing, including

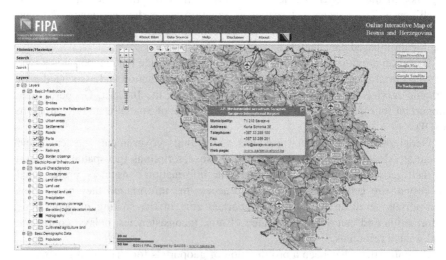

Fig. 3. Geoportal interface present data from spatial data warehouse

efficiency and cost effectiveness (reduce the cost of geospatial analysis by consolidating geospatial data marts, eliminating data redundancy and inconsistencies across applications, all of which leads to lower total cost of ownership); speed (speeds results by making data available for analysis sooner; there is no need to move data among systems); enhanced analytics (making better use of your location data enhances your analytics and BI); and scalability (can grow with business as data volumes increase).

3.2 The Interface

One of the most attractive and useful capabilities of geoportal is visualizing large amounts of information interactively. This ability to create multiple perspectives enhances a viewer's perceptive abilities to understand the phenomenon being studied.

Human-computer interface design focuses on how information is provided to and captured from users, and should provide a uniform structure for finding, viewing, and invoking the different components of a system. It actually defines the ways in which users interact with an information system. This is the reason we dedicated a large part of the work just for the geoportal interface, especially because it allows not only the interaction with the spatial data warehouse, but also present a kind of decision support system, as shown in Fig. 3.

4 Data Warehousing

Generally, a data warehouse is a large database designed to support the decision making needs of an organization as shown in Fig. 4. A data warehouse has been defined as a subject-oriented, integrated, time variant, nonvolatile collection of data that support a company's decision making process [9].

While data warehouses look at many types and dimensions of data, many are lacking in the spatial context of the data, such as an address, postal code, or provider location. By using technology that integrates this spatial component with the data warehouse, an organization can unlock hidden potential in their data and see hidden relationships and patterns in data, in essence data mining by geography.

In practice, there is evidence that spatially enabling database benefit an organization with more organized data structure; better integration of disparate data; new, spatially enabled analysis; reduced decision cycle time; and improved decisions. The spatial data warehouse extends the usefulness of online analytical processing (OLAP) systems. OLAP systems are used by decision makers to interrogate the data warehouse. The data for analysis with OLAP are accessed through metadata that document data source, frequency of update, and location of data. The data returned from the queries are represented as "multidimensional," although their form may be maintained as relational [10].

Spatial data warehouse, like Oracle BI, provides both a data model to the data warehouse and a geographic analysis engine for OLAP, which allows users to store spatial data inside the data warehouse [11]. It offers data transformation and manipulation, a spatial storage engine, robust data access mechanisms, and a broad range of analytical tools and methods that are designed to facilitate spatial analysis.

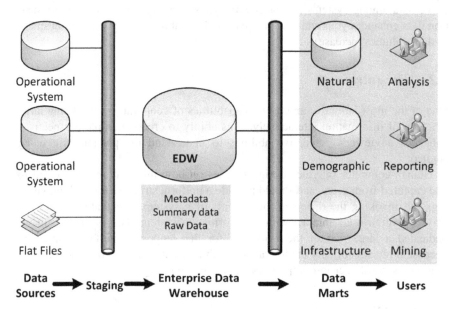

Fig. 4. Architecture of Business Intelligence System with Data Warehousing

5 Decision Support System with Spatial

Spatial data in BiH are characterized with fragmentation and lack of adequate availability of data, inconsistency, redundancy in collecting, insufficient use of standards, lack of coordination, and restrictions on data distribution. This situation makes it difficult to identify, access and use of existing spatial data in the country.

As per definition, this geoportal provides an entry point to access all data (geo-spatial data, remote sensing, information and services), and could be used to discovery, view, download, and transformation. We suggest building geoportal on three levels, web services platform, enterprise geoportal, and finally spatial data warehouse as shown in Fig. 5.

Web services platform contains Web GIS services, like Discovery service, means CSW (Catalogue Service Web), Viewing service, means WMS (Web Map Service), WCS (Web Coverage Service) and Download service, means WFS (Web Feature Service). Here also could be implemented geoprocessing services, open web services and tracking services. Enterprise geoportal contains catalog services, like Search, Channels, Link Browser Map, Download, and Collaboration. Finally, third level is data warehouse where data are stored.

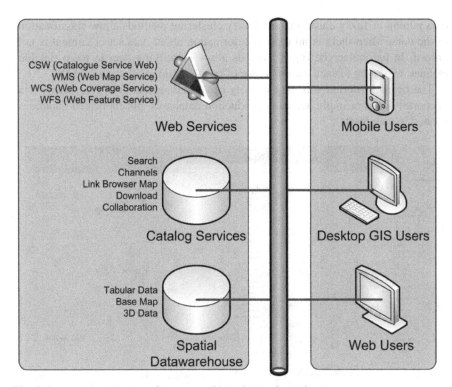

Fig. 5. Suggested architecture for geoportal based on web services

These services allow access to spatial information from different sources of local, national and global level in an interoperable manner and for a wider range of users to access relevant, harmonized and quality geographic information for the purposes of decision-making organizations and individuals.

We got information system that supports business decision making activities and serve the management and planning levels of organizations, show structural and non-structural data changing. Basically, this data warehouse represent knowledge based system, and we have interactive system intended to help decision makers compile useful information from a combination of raw data, documents, and personal knowledge, with their business models to identify and make decisions.

Summarizing previous statements we can see that it sounds like real decision support system. This decision support geoportal gather and present information like inventories of information assets (including legacy and relational data sources, cubes, data warehouses, and data marts); comparative statistic and demography figures between time points; and historic and projected economic indicators and natural characteristics based on statistic assumptions.

Generally, spatial-temporal domain is complex and characterized by a large amount of data as shown in Fig. 6. For advanced treat of data time series or huge amount data as spatial data usually are, there is need for advanced technique like data mining or fuzzy clustering are. Fuzzy clustering methods allow classification of the data, when there is no a priori information about data set or content is not known. In particular, the fuzzy methods allow identifying data in more flexible manner, assigning to each datum degree of membership to all classes. [12]

The design criterion was data security because distribution of information via geoportal does not imply access to production databases, but replicated databases located in data marts.

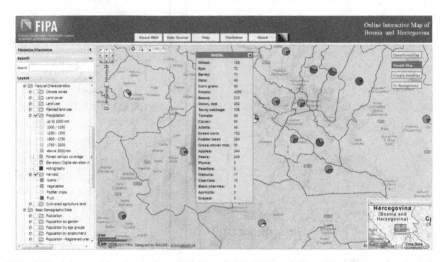

Fig. 6. Decision making in geoportal with available data from Data Warehouse

Implementation of the geoportal took following activities and deliveries: System analysis and design (design of system and software, as well as data models); Geoportal implementation and stuff training (policies and procedures for recovery, including failover functionality).

Authors recognized that there is need to research the extent to which open source software can support the development of geoportals as front-end of spatial data infrastructure especially in compliance with the INSPIRE directive [13].

5.1   Use Cases and Examples

There are more scenarios how users can make analysis and search for data in Online Interactive Map. Probably most popular methods are:

1. searching by mouse click on map
2. searching by Search form
3. predefined queries

Searches are initiated by clicking in the Search box and entering the search criteria. In most cases, these terms are "begins with" searches; meaning that user do not need to spell out the complete search criteria (Fig. 7.). As user entering search criteria, it initiates the search in real time and the results appear in the information pane immediately below the search criteria. In background of application is integrated complete functionality of searching technique based on metadata or on parts of the original texts represented in databases (such as attributes or locations).

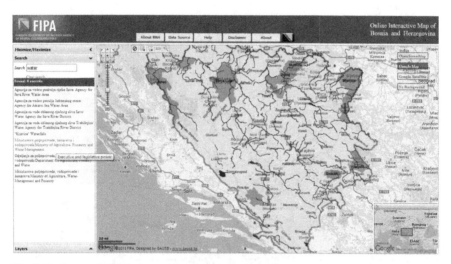

Fig. 7. Basic search for data

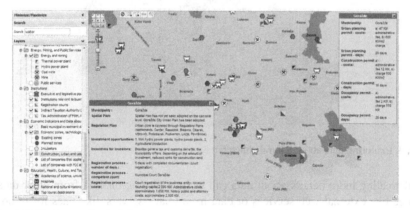

Fig. 8. Drill-down to basic infrastructure data and economic indicators

After that, potential investor can investigate Basic Infrastructure and Economic Indicator (investment data, economic zones, technological parks, incubators or construction, urban and use licenses) to better understand country potentials as shown in Fig. 8.

Using technique called "drill down" user goes from summary information to detailed data (Fig. 9.). In a GUI of GeoPortal, this means clicking on some representation in order to reveal more detail. It involves accessing information in database by starting with a general category and moving through the hierarchy: from category to table to record to field. Using this technique user performs data analysis on a parent attribute. It is a method of exploring multidimensional data by moving from one level of detail to the next one, where levels depend on the data granularity. Good example is retrieving data about natural characteristics.

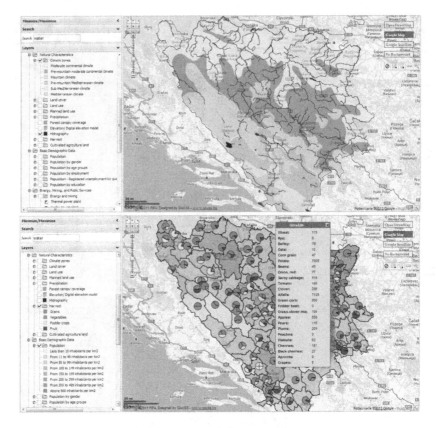

Fig. 9. Exploring data about natural characteristics

Revealing more details about data can include turning on data from other data source like external web services or data from image servers, like ones for orthophoto image presentation for whole country. This technique represent hybrid web visualization which enables integration different data format and from different sources. User can understand geography of location with measurements tools. Measurement panel displays distance, area, or radius depending on which tools in the toolbar are currently in use. Text appears in the following format: [label] [value] [unit of measure] to the right of the information panel. Example of this is presented in Figure 10.

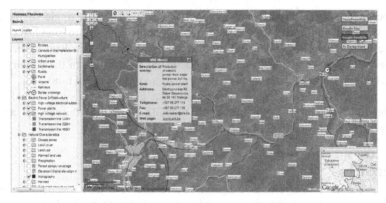

Fig. 10. Hybrid web representation with measurements functionality

At the end of analysis, user can summarize data, download in its own system and print it. Geoportal provides several map printing options. Since the current map extent shown in the active map window will be printed, generally user want first to zoom to the desired area and activate the desired layers before generating the printable output. This option allows generating printable maps in Acrobat PDF format. Also, this option offers more printing options, such as the ability to choose between several templates or scales. For investor most popular tool for final analysis is still Microsoft Excel, and they usually download analysis result in this format and continue with data analysis.

6   Conclusion and Future Work

The primary purpose of the geoportal is distribution and visualization of spatial data over the Internet, but its interactive capabilities could bring its functionality far beyond.

This paper proposed a usage of geoportals for decision making, especially with spatial data warehouses, whose main characteristics are: more organized data structure, better integration of disparate data, new spatially enabled analysis, reduced decision cycle time and improved decisions.

In the use case project, we utilized standard data warehousing infrastructure to integrate data from multiple source systems (different government agencies), enabling a central view across agencies. For agencies which could not accept the approach, we have created web services and catalogued all their data to central geo-portal.

We confirm advances in usage of spatial data warehouse to present spatial with business data together to generate thematic maps, because it present all information consistently, provide a single common data model, restructure the data so that it makes sense to the business users, and delivers excellent query performance, even for complex analytic queries.

One of the objectives in creating geoportal was to make it a single source for consistent spatial information. To achieve this goal there is need to establish cooperative mechanisms with other institutions and organizations involved in the collection, maintenance and distribution of geospatial information in the country, as well as monitoring and adjustment of the primary requirements of users.

As future work, we plan to incorporate case management system, add full collaboration system, and implement full WebGIS editing capabilities for agencies, who are interested to publish their data.

References

1. Tobler, W.: A computer movie simulating urban growth in the Detroit region. Economic Geography (1970)
2. Shekhar, S., Chawla, S.: Spatial Databases: A Tour. Prentice Hall (2003)
3. Worboys, M.F.: GIS: A computing perspective. Taylor and Francis (2004)
4. Batty, M., Longley, P.A.: Advanced Spatial Analysis: The CASA Book of GIS. ESRI Press (2003)
5. Goodchild, M.F., Fu, P., Rich, P.M.: Geographic information sharing: the case of the Geospatial One-Stop portal. Annals of the Association of American Geographers (2007)
6. Maguire, D.J., Longley, P.A.: The emergence of geoportals and their role in spatial data infrastructures. Computers, Environment and Urban Systems (2005)
7. Haining, R.: Spatial Data Analysis: Theory and Practice. Cambridge University Press (2003)
8. Abonyi, J., Feil, B.: Cluster Analysis for Data Mining and System Identification. Springer (2007)
9. Kimball, R., Ross, M.: The data warehouse toolkit: the complete guide to dimensional modeling. John Wiley & Sons, Inc. (2002)
10. Moss, L.T., Atre, S.: Business intelligence roadmap: the complete project lifecycle for decision-support applications. Addison-Wesley (2003)
11. Karabegovic, A., Ponjavic, M.: Integration and Interoperability of Spatial Data in Spatial Decision Support System Environment. In: MIPRO IEEE Croatia Conference, Opatija, Croatia (2010)
12. Karabegovic, A., Avdagic, Z., Ponjavic, M.: Fuzzy Clustering in Geospatial Analysis. In: MIPRO IEEE Croatia Conference, Opatija, Croatia (2011)
13. Petrovic, Z., Karabegovic, A., Ponjavic, M.: Spatial Data Infrastructure in Compliance with INSPIRE Directive and International Standards. In: 13th International Multidisciplinary Scientific GeoConference & SGEM 2013, Varna, Bulgaria (2013)

Prospects of Using Temporal Logics for Knowledge Management

Maria Mach-Król

Abstract. The paper concerns the possibility of using temporal logics for knowledge management. The idea of knowledge management is presented, along with the most typical computer solutions for this area. The temporal aspect of knowledge management is pointed out. Having in mind this temporal aspect, the paper presents possible advantages of extending knowledge representation for knowledge management with temporal formalisms.

Keywords: knowledge management, computer system, temporal logic, TAL language.

1   Introduction

Modern enterprises pay a lot of attention to the area of management that is called knowledge management. They understand, that employees' knowledge, or more generally speaking, the knowledge of organization, constitutes one of its key resources. Therefore basic management trends encompass not only managing quality or change, but also knowledge management. It is this area of activity that enables an enterprise to compete with its competitors on the more and more turbulent and dynamic markets.

It must be noted, at the same time, that the most of knowledge is of temporal character. Knowledge changes in time – for two basic reasons. The first is simply the flow of time, while the second – gathering of new information about objects, that knowledge concerns, objects that possess temporal characteristics [2]. Therefore omitting of a temporal dimension would lead to loosing of important knowledge elements. In this way, time becomes an important category for an enterprise in the area of knowledge management.

While analyzing current informatics solutions for knowledge management, it has to be noticed that time as a knowledge dimension is not noticed at all. Taking into account importance of a temporal aspect, it seems a major disadvantage. Therefore in this paper we propose extending of a knowledge representation in

Maria Mach-Król
University of Economics, Katowice, Poland
e-mail: maria.mach-krol@ue.katowice.pl

M. Mach-Król and T. Pełech-Pilichowski (eds.), *Advances in Business ICT*,
Advances in Intelligent Systems and Computing 257,
DOI: 10.1007/978-3-319-03677-9_3, © Springer International Publishing Switzerland 2014

knowledge management systems by temporal formalization, and we consider advantages of the proposed solution.

The paper is organized as follows. In Section 2 the concept of knowledge management is presented. Section 3 concerns computer solutions in this area. The next point (Section 4) is devoted to the temporal aspect of knowledge, and to the advantages of using temporal formalization. In Section 5 a practical example of temporal knowledge management is presented. The last section of the paper contains conclusions.

2   Concepts and Models of Knowledge Management

Nowadays knowledge is perceived by modern enterprises as one of key resources that is equally (or more) important as such "classical" types of resources as land, capital or work. What make knowledge so important are its features. M. Grudzewski and I. Hejduk ([5] s. 48) point out the following knowledge's features:

- Domination – meaning, that knowledge is the most important resource of a firm;
- Inexhaustibleness – knowledge that is used, spread, moved does not diminish;
- While used, knowledge gathers value, not used, it disappears;
- Simultaneousity – knowledge may be used by many persons at the same time;
- Non-linearity – it is not possible to point out a direct relationship between the amount of knowledge possessed and the advantages of it.

The above mentioned knowledge features created (among other features) the management trend called knowledge management, because knowledge role as a resource has been noticed. It is a relatively young domain in management sciences, therefore does not exist a commonly accepted definition of knowledge management. The authors of [5], cited before, assume knowledge management as "the whole of processes enabling creating, spreading, and using knowledge for organization's purposes" (p. 47). This definitions links explicitly to the temporal dimension of knowledge, because it uses a definition of processes, which is linked with change. Modeling of processes is useful while describing continuous phenomena, as economic reality for example, therefore it is also useful for describing changes of knowledge treated as enterprise's resource. More on this topic may be found in [17].

Definitions of knowledge management are numerous, as are models of knowledge management. In the literature, the most important models are: the resource one, the Japanese one, and the process one.

The first one – the resource model – treats knowledge as a key resource of an enterprise. This resource comes both from the inside of an organization, as from its environment. In this model, the purpose of an enterprise is getting the strategic competitive advantage in the area of knowledge resource and its usefulness. More on this topic may be found in [3], and [16].

The name of the Japanese model comes from the nationality of its creators (I. Nonaka, and H. Takeuchi). They formalized Japanese firms' experience. The main accent in the model is put on knowledge creation.

Finally, a process model. It is based on previously cited definition of knowledge management as a set of processes, of which the most important are gathering and creating of knowledge, knowledge dividing, and transforming knowledge into decisions. The temporal character of the model (given *implicitly*), linked with the process description, has to be stressed here once more.

3   Computer Apparatus for Knowledge Management

Although in some definitions of knowledge management we may find some links to its temporal aspect (see Section 1), and although temporal dimension is present also in the definition of knowledge management system (see below), these systems do not possess ability to represent temporality explicitly.

As the author of [16] points out, knowledge management systems are "information systems that help workers in an enterprise with performing processes linked with knowledge management, such as location and acquisition of knowledge, its transfer, development and use" (p. 54). In the work cited, a schema of computer knowledge management system can also be found. It is presented in Fig. 1.

For the purposes of this paper, the application layer is meaningful, that is the layer consisting of knowledge management computer tools. On a general level, one can point out such systems, as ERP, CMS or search engines, while on a more detailed level, computer tools for knowledge management encompass for example:

- Document management systems,
- Competences management systems,
- Community management systems,
- Workflow systems
- Content management systems,
- E-learning systems,
- Searching systems,
- Groupware systems.

Applications for knowledge management are shown in Fig. 2; they will be also presented later on in a more detailed manner.

According to the elements of Fig. 2, the most popular and typical knowledge management systems are as follows:

- Documents management systems – which create, classify, create electronic archives of documents;
- Competences management systems – they create, write, publish, plan and analyze employees' competences;

- Workflow systems – automate processes of passing information, documents or tasks from one employee to another, according to a timetable;
- Community management systems – enable members of a "community" to communicate, where a "community" may be a group or groups of employees working on the same project;
- Content management systems – where "content" is understood as contents of web pages, intranets, multimedia etc.;

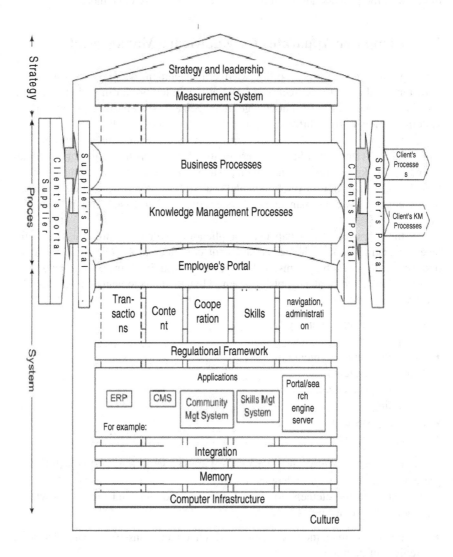

Fig. 1. Architecture of a knowledge management system. Source: [16] p. 55.

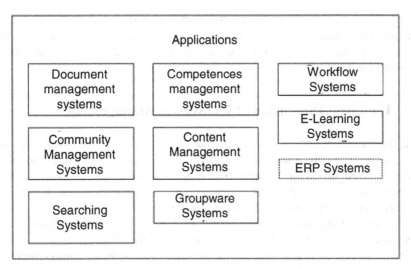

Fig. 2. Application layer of a knowledge management system. Source: [16] p. 56.

- E-learning systems – allow learning with the use of internet. Therefore their aim is to create and diffuse knowledge;
- Searching systems – aimed at a specific kind of classification, concerning search of documents, search of information inside documents, search of metadata on documents. Nowadays space the most frequently searched is www;
- Groupware systems – this term concerns software enabling exchange of information between members of the group working on the same task. It also enables – among others – planning of meetings (time management) or contacts management;
- ERP systems – systems that succor management in enterprises and institutions, with economic and planning functions. They enable to optimize internal and external processes of an enterprise. They encompass planning of all assets of an enterprise; therefore also knowledge perceived and treated as an asset.

As the above short survey of tools has shown, no one of them has implemented explicitly a possibility to handle temporal dimension of discourse. Some elements of activities linked with the notion of time are of course present. For example, archiving documents allows for tracing their changes, competence planning (e.g. training plan) is also settled in time, as well as task planning in groupware systems. It must be said that it is nevertheless the simplest kind of temporal dimension, linked with calendar time axis. No more advanced mechanisms can be found, that would enable for example analysis of reasons for knowledge changes, tracing knowledge evolution etc. Such possibilities are offered by systems based on temporal logics, which can perform temporal reasoning in an explicit and direct way (more on this topic can be found e.g. in [10]). It seems therefore that incorporating temporal formalisms into existing systems, or constructing new, fully temporal tools would constitute a great extension of possibilities in knowledge management. The next Section presents advantages of using temporal logics and formalisms.

4   Temporal Dimension of Knowledge Management

As it has been already pointed out in the Introduction, the knowledge in an organization is mostly temporal in characteristics. This means, that with the passing of time knowledge changes, new information comes on objects, that knowledge concerns, if these object poses temporal characteristics. It can be therefore said that this knowledge dimension, that is called "time" is in this case explicit. So omitting this dimension would lead to losing important elements of knowledge – temporal features. Having this in mind, time becomes for an enterprise a very important category in the area of knowledge management. It seems that enriching at least some of knowledge management systems with the possibility of explicit expression of temporal knowledge aspect would allow bettering managing this knowledge, even if taking into account its dynamics. The basic way of representing the temporal aspect of any phenomenon, including knowledge, is the use of temporal logics. Using this group of logic formalisms for knowledge management would lead to several advantages, coming from the advantages of temporal representation. Using temporal representation is well motivated, there are a lot of theoretic works on temporal formalisms and their features, also temporal formalisms have been used in many domains. It is certain, that temporal representation of a domain – including organizational knowledge – has many advantages. They can be divided into several groups:

1. Basic advantages – concerning temporal representation itself, independently from where it is used; these basic advantages also are the origin of advantages from other groups;
2. Advantages concerning representation of change;
3. Advantages concerning representation of causal relationships.

Time, as a dimension, is a basis for reasoning about action and change – only a proper use of temporal dimension allows for representation of change and its features, as e.g. its scope or interactions caused by change [15]. Such explicit temporal reference is possible through the use of a temporal formalism, where time is a basic variable.

Temporal logic allows encoding both qualitative and quantitative temporal information, as well as relationships among events, therefore it is easy to express such relations, as "shorter", "longer", "simultaneously", "earlier" etc. This in turn implies easiness of arranging phenomena in time, even if they overlap – Allen's interval algebra is an example of a formalism which allows such arrangements.

Temporal formalization makes possible to encode discrete and dense changes (according to a model of time adopted), allows for describing change as a process, and for reasoning about causes, effects and directions of change.

As time is the fourth dimension of the world, it may not be omitted during the reasoning process; otherwise the perspective of analysis would be too narrowed. The temporal dimension allows the system to "learn": the system collects cases concerning a phenomenon (or a subject domain) being represented, traces its evolution and thanks to this is able to generate new solutions [7].

It has been already said that temporal representation makes possible to represent change as a process. It is so, because with temporal logic, processes can be modeled explicitly – therefore knowledge on their temporal aspect, their interactions, on concurrent processes is easily expressed [1]. As Kania points out ([17], p. 60), models of processes are useful for describing dense phenomena, as for example economic ones.

Temporal logic gives us richer – temporal aspect included – formalization of domain knowledge, it also gives us "knowledge on knowledge": combining temporal operators with formal knowledge representation one can formulate assertions about knowledge evolution in a system [6]. Van Benthem presents an example of such combination, suggesting combining temporal and epistemic logic [2], p. 335. Placing knowledge in time treated as a basic dimension, one can add new knowledge to a base, not removing the "old" one, and with no risk of inconsistencies [8]. Temporal logic, as a knowledge representation language, should provide both explicit knowledge and access to tacit one ([12], p. 326). Temporal logic, which has reasoning rules built in, is able to provide this property.

Summing up, it should be pointed out that temporal formalisms meet the requirements of knowledge representation in artificial intelligence, such as:

- expressing imprecise and unsure knowledge,
- expressing "relations" of knowledge (e.g. A occurred before B", that very often have no explicit dates;
- different reasoning granulations,
- modeling of persistence.

The above postulates are met e.g. by Allen's interval algebra [1]. Therefore enriching the existing knowledge management systems with temporal formalisms, or building new systems, based on these formalisms, would allow for taking into account the temporal dimension of knowledge, its changes and evolution/development. In this way knowledge may be managed more effectively.

5   Example of Temporal Knowledge Management

We will present a practical example of temporal knowledge management, concerning a problem of establishing, whether an unemployed person may be granted a benefit. The example chosen is very simple, to focus attention on a temporal languages approach. The temporal language chosen is TAL.

The TAL (Temporal Action Language) is derived from Sandewall's PMON logic [14]. Its main features as a language for describing temporal dependencies include: the notion of time independent from actions, the possibility of defining causal dependencies apart from actions' definitions, and the possibility of describing concurrent interactions.

The language consists of two levels (layers): the so-called surface language, which is used to describe narratives (for more information see [4], and the so-called base language, namely the logic of events, which is an ordered 1st order predicate logic. Any correct narrative description, after being transformed into the description in the base language constitutes a finite set of 1st order wffs.

The surface language layer consists of:

- temporal expressions,
- value expressions,
- atomic expressions,
- narrative statements,
- additional macro-operators and abbreviations (e.g. the durational reassignment operator, the reassignment operator, the occlusion operator etc.).

The base language layer (the logic of events) contains, among others, temporal predicates: HOLDS, OCCURS, OBSERVE, DUR, PER and others (the exact definitions of the predicates can be found e.g. in [4].

The surface language does not have formal semantics, although it has a formal syntax. The whole formal inference process is conducted after "translating" the description in the surface language into the description in the base language.

It is also worth mentioning here, that the description (specification) of a scenario in the TAL language consists of:

- type description,
- action definitions and descriptions,
- domain constraints specification,
- temporal dependencies specification.

To implement solutions encoded in the TAL language, one can choose the VITAL tool [9], developed at Linköping University, Sweden. The tool is very easy to use and automatically translates scenarios from the surface language (TAL) into the base language, therefore the person constructing a scenario does not have to be an expert in temporal predicate logic.

The example will be illustrated with Canadian unemployment law. The illustration comes from [13] and [11]. The authors present there a concrete decision problem, which in our opinion could be solved by using the TAL language. All the rules come directly, or after slight modifications, from Canadian unemployment law. The example to be discussed is presented in Fig. 3.

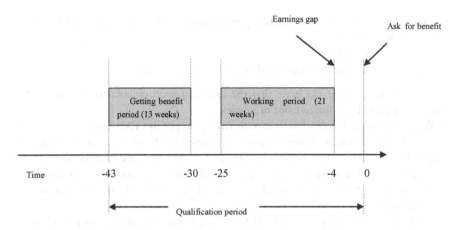

Fig. 3. Sample decision situation. Source: own elaboration based on [13].

As we can see in Fig. 3, a person asking for benefit lost work (in a week numbered with -4) and asked for benefit in a week numbered with 0. The former working period lasted for 21 weeks. The person had already been granted a benefit, and has been getting it for 13 weeks. For our purposes it is not important, what happened between weeks -25 and -30. We may assume that it was a working period or an unemployment period. The most important information concerning working period is that it lasted for minimum 21 weeks: on this basis we may establish, whether the person qualifies for a new benefit (this results from the Canadian law).

The basic problem (as in original works [13] and [11]) is to establish, whether the person has the right to the new benefit period. In our example the answer is yes, because the former working period was longer than 20 weeks (more details about the legal rules are to be found in [13], [11]. It is apparently a temporal information. Next we have to establish, how long a new benefit period is to be. This in turn depends on the information about the former benefit period, because the starting point of so-called qualification period is the point in which the former benefit period started. As it can be seen in the figure, the qualification period for the sample person is 43 weeks.

Summing up – the temporal aspect of the problem concerns establishing the longitude of qualification period and – in consequence – the new benefit period of an unemployed person.

The tool for the implementation of the problem is – as said before – the TAL language and its implementation named VITAL. The main task for VITAL was to calculate the longitude of qualification period, needed for establishing, whether a person will be granted a benefit, or not. As input data the following information has been provided:

- The fact, that the person has already been granted a benefit in the past (and for how many weeks),
- The former working period,
- A point time in which a person asked for the new benefit.

It should be pointed out here, that the assumed time granularity is one week, therefore time point is the number of a week.

We have encoded input data in a scenario written in the TAL language, in form of so-called occurrences. From these occurrences the VITAL tool was to infer the longitude of a qualification period. To enable this inference, we had also to encode a proper temporal dependency, linking input facts with the way of calculating the qualification period. The dependency has been taken from the Canadian unemployment law.

A key to success (that is to establish the longitude of qualification period) is a rule stating, that qualification period starts in the same time point as the former benefit period, and ends in the time point in which a person asks for a new benefit. Therefore, in terms of the TAL scenario, if we denote by *t1* the time point in which the variable *getting_benefit_1* becomes true, and by *t2* – the time point in which action *ask* (asking for new benefit) becomes active, then the variable *qual_period* (denoting the qualification period) should become true (the system should infer this value) over the interval [t1, t2). The discussed dependency, has the following form in the TAL language:

```
dep forall t1, t2 [
    Ct([t1] getting_benefit_1) &
    Ct([t2] ask) &
          !exists t3 [t1 <= t3 & t3 < t2 & Ct([t3]
ask) ] ->
          I([t1, t2) qual_period) ]
```

where:
dep – dependency label,
t1, t2 – time points,
Ct – operator *changes to true*,
I – macrooperator, that assigns a new value to the default value of a feature, over a particular time interval ([4]).

The dependency stated as above will work also in a situation, when the same person will be hired again, then will get benefit again (*getting_benefit_1*), and then will again ask for benefit (*ask*). In such a situation the variable *qual_period* will be true from the first moment in which *getting_benefit_1* becomes true, up till the last action of asking for benefit. By formulating the dependency in the form presented above, we may check, whether *t2* is the shortest time point after *t1* such, that Ct([t2] ask).

The sample situation presented in Fig. 3 has been encoded in the VITAL tool. The only difference was that while numbering time points (weeks), positive numbers have been used. The result of temporal reasoning performed by the VITAL tool is presented in Fig. 4.

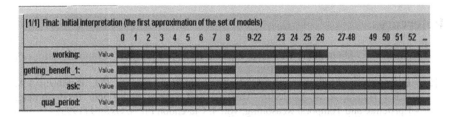

Fig. 4. The result of calculating the qualification period in the VITAL application. The lighter color on the timeline denotes the true value of a feature, while the darker one denotes the false value (in reality the drawings prepared by the VITAL tool are colorful). Source: own elaboration. Working - the time person has been working, other notations as in the text of the paragraph.

What is known after performing temporal reasoning by the VITAL tool? It is known for how many weeks the person has been getting the former benefit (if at all; in our example – for 13 weeks), for how long the person has been employed during the qualification period (from this information it will be possible to infer if the person may be granted the next benefit; in our example the working period is 21 weeks), how long is the qualification period (in our example – 43 weeks). All this information can be read from the above figure.

Moreover, we may assume that while asking for benefit, a person is obliged to provide some additional information, e.g. on family situation, education, additional skills etc., to enable decision on other forms of help. It makes no sense encoding this information in the VITAL tool, as it is not of temporal kind. Nevertheless the results of reasoning performed on temporal data, together with the additional information, may constitute an input for further, non-temporal reasoning.

6   Conclusions

Knowledge management is nowadays one of the most intensively developing trends in management. It is so because the growing role of knowledge in economic success and competitiveness is noticed and appreciated. At the same time it is important, that knowledge is mostly temporal in nature: knowledge changes in time. Therefore the temporal aspect of knowledge may not be omitted while managing this important asset of an enterprise.

In the existing computer systems for knowledge management the temporal aspect is present rarely and implicitly. Taking into account its importance, in the paper we proposed to use temporal logics to extend functionality of existing systems, or to build new computer tools for KM. It seems that the advantages of using temporal formalisms, presented in the paper, make this postulate fully justified.

References

[1] Allen, J.F.: Maintaining Knowledge about Temporal Intervals. Communications of the ACM 26(11) (November 1983)
[2] Benthem van, J.: Temporal Logic. In: Gabbay, D.M., Hogger, C.J., Robinson, J.A. (eds.) Handbook of Logic in Artificial Intelligence and Logic Programming. Epistemic and Temporal Reasoning, vol. 4. Clarendon Press, Oxford (1995)
[3] Bielecki, W.: Wirtualizacja Nauczania In: Wawrzyniak B., Zarządzanie Wiedzą w Przedsiębiorstwie, Warszawa, p. 317 (2003)
[4] Doherty, P., Gustafsson, J., Karlsson, L., Kvarnström, J.: Temporal Action Logics (TAL): Language Specification and Tutorial. Linköping Electronic Articles in Computer and Information Science 3, 015 (1998), http://www.ep.liu.se/ea/cis/1998/015
[5] Grudzewski, W., Hejduk, I.: Zarządzanie Wiedzą w Organizacjach. E-mentor 8, 46–51 (2005)
[6] Halpern, J.Y.: Reasoning About Knowledge: A Survey. In: Gabbay, D.M., Hogger, C.J., Robinson, J.A. (eds.) Epistemic and Temporal Reasoning, vol. 4. Clarendon Press, Oxford (1995)
[7] Jakubczyc, J.: Wprowadzenie do ekonometrii dynamicznej. Wydawnictwo Naukowe PWN, Warszawa-Wrocław (1996)
[8] Kowalski, R., Sergot, M.: A logic-based calculus of events. In: Schmidt, J.W., Thanos, C. (eds.) Foundations of Knowledge Base Management: Contributions from Logic, Databases, and Artificial Intelligence, pp. 23–55. Springer, Heidelberg (1989)
[9] Kvarnström, J., Doherty, P.: VITAL research tool (1997), http://anton.ida.liu.se/vital/vital.html
[10] Mach, M.A.: Temporalna analiza otoczenia przedsiębiorstwa. Techniki i narzędzia inteligentne. Wydawnictwo AE Wrocław (2007)
[11] Mackaay, E., Poulin, D., Frémont, J., Bratley, P., Deniger, C.: The logic of time in law and legal expert systems. Ratio Juris 3(2), 254–271 (1990)
[12] Mylopoulos, J., Borgida, A., Jarke, M., Koubarakis, M.: Telos: Representing Knowledge About Information Systems. ACM Transactions on Information Systems 8(4), 325–362 (1990)
[13] Poulin, D., Mackaay, E., Bratley, P., Frémont, J.: Time Server – a legal time specialist. In: Proc. Third International Conference on Logic, Informatics, Law, Florence, November 2-5 (1989)
[14] Sandewall, E.: Features and Fluents. Oxford University Press (1994)
[15] Vila, L.: A Survey on Temporal Reasoning in Artificial Intelligence. AI Communications 7(1), 4–28 (1994)
[16] Wykowska, G.: Techniki informacyjne w zarządzaniu wiedzą. MSc Dissertation, SGH Warszawa (2008)
[17] Kania, K.: Temporalne bazy danych w systemach informatycznych zarządzania. Prace Naukowe Akademii Ekonomicznej im. Karola Adamieckiego w Katowicach, Wydawnictwo AE Katowice (2004)

On Problems of Automatic Legal Texts Processing and Information Acquiring from Normative Acts

Tomasz Pełech-Pilichowski, Wojciech Cyrul, and Piotr Potiopa

Abstract. In the paper, problems of legal information digitalization are investigated. Conditions for extraction information from legal texts (i.a. normative acts) related to the common ones processing (non-legal terms, in English) are outlined. Problems of dimensionality reduction and application of similarity measures are discussed. Sample results of similarity analysis is presented. Further research aimed at semantic analysis of legal texts are outlined.

1 Introduction

Theoretical and practical problems concerning digitalization of legal information integrates contemporary legal theory and legal informatics. Since the 1960's, several projects have been lunched aiming at the computerization of legislative resources, and making them more easily accessible for users. Some of them i.e., JURIS[1], ITALGIURE-FIND[2] or Noris are still running, while others, like FLITE,[3] ended [22]. The Internet had brought a new dynamic to this process in the second half of the 1990's. It opened the theoretical and technical possibility to promulgate and communicate legal texts in a machine-readable form. It also promised practical realization of the idea of more efficient and more transparent legislation, available on-line for lawyers, business people and eventually for all citizens. Since that time, IT and ICT start to play an ever-growing role, both in legal discourse and in legislative processes. However, the use of IT in law raises numerous new problems. Most important ones are: a question of standards used in the process of

Tomasz Pełech-Pilichowski · Piotr Potiopa
AGH University of Science and Technology, Al. Mickiewicza 30, 30-059 Krakow, Poland
e-mail: tomek@agh.edu.pl, ppotiopa@zarz.agh.edu.pl

Wojciech Cyrul
Jagiellonian University, ul. Gołębia 24, 31-007 Krakow, Poland
e-mail: cyrulwojciech@hotmail.com

[1] http://www.gesetze-im-internet.de
[2] http://www.italgiure.giustizia.it
[3] Federal Legal Information Trough Electronics, see ABA Journal February 1976, vol 62, What's New in the Law, pp. 233-7, A. Ashman

M. Mach-Król and T. Pełech-Pilichowski (eds.), *Advances in Business ICT*,
Advances in Intelligent Systems and Computing 257,
DOI: 10.1007/978-3-319-03677-9_4, © Springer International Publishing Switzerland 2014

digitalization of normative acts or other legal documents; finding a method of adding computer-processed information to these texts; and finding tools which will help support both their preparation and management in a standard compliant way [27].

The potential of IT and ICT was also very soon recognized and acknowledged in Poland. In fact, the debate on the complexity of digitalization of law had already started in the end of the 1960's. From the beginning, it had a pragmatic character and focused mainly on possible applications of IT and ICT in law (see [25], [11], [16], [23]). This instrumental approach became even more pervasive in the late 1990's when the discussion began to be dominated by the intricacies related to building and functioning of legal databases and complications connected with regulation of the computerization of law ([28]). As a result of this development, the Act *of June 20, 2000, on promulgation of normative acts and some other legal acts* (Dz. U. Nr 62, poz. 718) states that official legal journals and collections of local law should be issued and stored not only in the printed but also in the electronic form[4]. Not much later, the statute is amended by the *Act of March 4, 2011 r, on amendment of the statute on promulgation of normative acts and some other legal acts* (Dz.U. Nr 177, poz. 676). In consequence, since January 1, 2012, Polish law has been published exclusively in an electronic form and communicated on Internet, like in several other EU countries i.e. Austria, Belgium Denmark, Spain, Portugal or Hungary[5]. It should be remembered however, that these change were preceded both by development of several technical standards and by extensive regulation how these standards should be implemented in law.

The legal framework concerning technical standards for electronic legislation in Poland is constituted currently by the *Act of February 17, 2005 on computerization of the activities of the entities implementing public tasks* (Dz.U. 2005 nr 64 poz. 565) and the *Regulation of the Prime Minister of December 27, 2011 on the technical requirements for electronic documents, which contain normative acts and other legal acts, for official journals published in the electronic form and for means of electronic communication and electronic data carriers* (Dz.U. Nr 289, Poz. 1699). Regretfully, existing regulations focus more on security and integrity of electronic legal documents than on increasing effectiveness of managing information in the legislative process and enhancing cooperation between different legal and social agents contributing to the process. Moreover, electronic legal documents containing normative acts and other legal acts announced on the basis of the abovementioned *Act of July 20, 2000*, shall be prepared as structural texts in a XML 1.0 format, as defined by W3C. The XDS schemes of such documents, although being developed and published by the government, have only the status of recommendations, and thus are legally not binding. In other words, although,

---

[4] See articles 20 a paragraph 1 and article 20 b.

[5] See Access to Legislation in Europe 2009Guide to the legal gazettes and other official information sources in the European Union and the European Free Trade Association, p.229 [http://circa.europa.eu/irc/opoce/ojf/info/data/prod/data/pdf/AccessToLegislationInEuropeGUIDE2009.pdf]

according to the law, electronic documents containing legal texts should stay in line with the patterns of legal acts published in central repository of patterns of electronic documents and on the web page *www.dziennikiurzedowe.gov.pl,* the patterns as such do not need to be used when drafting a legal text[6]. The law requires only that the structure of a legal act, defined in XML, should be in accordance with the structure of legal acts as determined in the *Regulation of the Prime Minister of June 20, 2002 on legislative technique* (Dz. U. Nr 100, poz. 908) and promulgated in a PDF format or in a XML format enabling visualization of the content in the form of a PDF file[7]. This situation creates a risk that electronic documents containing legal texts will not be fully compatible, and therefore hinders effective management and processing legal information in future. It poses also a significant problem, for the reason that the information included in the legal resource is complex and heteronymous. Effective and appropriated interconnection of its different layers requires not only a multi-layered architecture for modeling legal documents but also existence and implementation of shared and open standards for legal knowledge representation (see [19]).

The goal of the paper is to analyze a possibility of performing reliable similarity analysis of sample polish legal acts, based on approaches and software tools typically adopted to common language processing (especially sentences, terms in English). The primary motivation is to show the results of the preliminary work in the context of long-term research. General problems of legal information digitalization are investigated. Conditions for extraction information from legal texts related to the common ones processing (non-legal terms) are outlined. Selected steps, including dimensionality reduction and application of similarity measures, are discussed. In addition, semantic similarity analysis based on WordNET is outlined. Sample results of similarity analysis focused on distinguishing groups of legal acts from similar semantic area are presented (for this purpose, dedicated application for reading legal texts available on-line as PDFs was prepared). Further research aimed at semantic analysis of legal texts are outlined.

2   Digitalization of Legal Information

The ongoing digitalization of legal information and gradual moving the traditional activities of the state to the cyberspace result mainly from the states' needs to increase efficiency of legal drafting and administration of justice. However, the process is also supported by the growth of e-commerce and the ever-present expectation of lowering costs of legal transactions. This situation changes not only the way, in which law is being made and applied but also alter the ways in which

---

[6] Currently the XML schemas (xsd) of polish legal acts is developer one the electronic platform of public administration services (ePUAP). The latest recommendation in this respect was published in February 2012.
http://epuap.gov.pl/wps/PA_1_FCHRSKG108KP5023L0FQM82083/prod ukt_rekomendacji/produkt_rekomendacji_340

[7] §8.1 of the Regulation.

legal information is searched, aggregated, analyzed stored, erased and communicated. Respectively, new standardized forms of legal documents are being formed, new forms of notification and delivering of legal decisions in legal proceedings are being developed, new tools are being used to encode and to manage legal information, and new electronic public archives or registers are being created.

Abovementioned developments open up an important field of activity for informatics. It is especially the case because to manage effectively legal information in the digital form, normative texts and various types of legal documents must be drafted and disseminated with the use of computers. Furthermore, the special status of legal information requires development and implementation of standards, which secure both its authenticity and integrity. However, this is not at all an easy task, taking into account political dimension of lawmaking, the increasingly international and decentralized characteristic of both legislative and juridical activities and the growing diversity of agents providing legal information.

The effective accessibility, security and certainty of legal information on the Internet require implementation of shared machine-readable standards and the development of technical solutions, which will enable the interoperability of various systems and tools. Furthermore, these standards must take into account different aspects and particularities of legal texts, i.e. the appearance, the structure or the different normative status of particular types of legal texts[8]. Although, these standards are being developed and gradually implemented, also in Poland, there is surprisingly little discussion about what standards should be applied, or how existing standards should be modified in order to fit best into the present and the future needs of both Polish state and the addressees of law. More discussion is also needed with respect to both the potential problems resulting from the specific semantic, syntactic and pragmatic nuances of legal texts and the impact of implemented technologies on legal practices. One must remember that status, types, structures and relations between different legal texts are determined by law. Furthermore, while some links or relations between different legal provisions are clearly defined by law others are identifiable only on the conceptual level. Taking it all that into account, there is no doubt that the standards implemented today for digitalization of legal information will determine the possibility, scope, and costs of further developments. The awareness of these factors is important also because new standards are still being developed, aiming to meet new expectations of business people and lawyers. For example, there is a growing need for enabling an online access to legal information with the use of different electronic devices and for development of tools allowing computerization of aggregation of legal information and its analysis, taking into consideration the semantic dimension of law.

Both, the scope of digitalization and computerization of legal information depend on existence and quality of so called *knowledge representation languages.*

---

[8] One must remember also that legal documents tend to be much longer than usual documents searched on-line, they exhibit a wide range of internal structures, have a significant amount of editorial value added, they have particular normative force, contains non text elements and they play different role in social discourse. See more: [26]

They permit a description of legal information in a way that enables the machines its application. Eventually, information systems could not only support actions of individuals through versioning, finding or evaluating particular information, but also apply directly and automatically legal provisions. To some extent, it is already possible since any graphical sign can be represented in an alphanumerical way (i.e. ASCI or Unicode). There are also naming convention permitting univocal identifications of all available online legal documents or documents to which legal documents directly refer to (i.e. URL or URI). Furthermore, thanks to XML markup language, a machine-readable representation of the formal structure of legal documents is also possible. Unfortunately, there are still problems with making the machines "understand" and apply legal texts, since there are neither universal legal ontology nor the universally acceptable model of legal reasoning. The already existing tools have either a limited scope of application or are unable to computerize legal practice to the extent which would not create any doubts about the validity of conclusions.

Taking the above into account, one can see that efficiency and the scope of computerization of legal information depends, to large extent, on the existence of shared standards and conventions enabling the technology neutral representation of legal information. Standardization could increase the efficiency of generation, presentation, accessibility and description of legal documents. Furthermore, an existence of shared standard for legal information significantly reduces costs of digitalization of legal information and guaranties higher level of interoperability of different systems. In fact, it can also help to monitor a consistency of national regulation and more effectively compare legal institutions belonging to different legal systems. Most importantly however, it facilitates the development of variety of tools used for management of legal text, such as editors, converters, name resolvers, validations tools, search engines, workflow managers, intelligent agents or publishing systems (see [27]). It also promote co-operation between different agents, both on national and international level, facilitating knowledge dissemination and better coordination of mutual actions.

3 Morpho-Syntactic Text Analysis

3.1 Processing of Polish Legal Texts: Steps and Tools

Discussed in previous chapters, agreed by law, standards of legal texts formats are the great opportunity to create tools able to search legal information and to build semantic relations between the particular sets of texts. Therefore, considering efficient information analysis and extraction with automated tools, the following areas of obtaining information seem to be vital to investigate:

- searching keyword phrases in legal texts and locating them in selected blocks of legal texts,
- automatically examining relations between legal texts (such as similarity),

- the law thematic area recognition on the basis of the legal test samples,
- legal texts semantic analysis and (semi) automatic legal ontology construction.

As mentioned above (see Introduction), adopted in the Polish law regulations on the media publication of legal texts say that the format adopted in the generation and publication of these texts is XML and PDF (for publication of texts on the Internet). Developed XML is compatible with the developed schema XSD file approved and published[9]. A PDF file may be generated based on XML thus further considerations mainly concern XML file processing. Therefore, it allows to assume that the processing of legal acts published on the Internet[10] will be sufficient.

Unfortunately, despite our attempts to find laws generated in XML format adopted in accordance with established XSD schema, we could not find such documents. Therefore, it became more difficult to extract information and document fields for research, however, such information simplifies the similarity analysis as:

- structured way of writing a legal text eliminates the problem of syntactic ambiguity in analysis and parsing of the document;
- structured schema makes possible to clearly separate systematization and editorials units of the legal text clearly defined in schema, respectively: "*część*" ("part"), "*księga*" ("book"), "*tytuł*" ("title"), "*dział*" ("section"), "*rozdział*" ("chapter"), "*oddział*", *artykuł* ("article"), "*paragraf*" ("paragraph"), "*ustęp*", "*punkt*" ("point"), "*litera*" ("letter"), "*tiret*";
- facilitates the validation of the document with the adopted scheme XSD and use of tools for parsing XML documents[11].

3.2   Term-Document Matrix

3.2.1   TF-IDF Weighting

To achieve better description of dependencies between analyzed documents, the modification of the TDM matrix basic structure has been employed. TDM (term-document matrix) consists of words extracted from the input data. TF-IDF factor is based on determining the relative frequency of a term and then comparing to the inverted frequency of term calculated in the entire collection of documents.

For each term, TF (term frequency) is the relative frequency of the term w occurrences in the document d (thus the term rank) and the IDF (inverse document

---

[9] The electronic form of legal acts – Legal Acts Editor – EDAP
http://bip.msw.gov.pl/portal/bip/185/18658/Edytor_Aktow_
Prawnych_EDAP__wersja_z_dnia_31_marca_2010_r.html
(access: 12.07.2013)1

[10] For example, see http://isap.sejm.gov.pl

[11] In the testing application, described in the article we used a module that is Java DOM Parser.

frequency) as inversely proportional to the occurrences of the term in relation to the corpus D ($d \in D$) of documents, which represents an importance of the word in the entire collection of documents. To calculate TF-ID factor is calculated as follows (eq. 1) [21], [20]:

$$w_d = f_{w,d} \, log\big(|D|/f_{w,D}\big) \tag{1}$$

where $f_{w,d}$ is a number of occurrences of w in d, $|D|$ is the size of document corpus $f_{w,D}$ is a number of documents containing term w.

For large collections of documents, the normalization of the TF with dimensionality reduction procedures may be used.

3.2.2 Dimensionality Reduction

An availability of large datasets makes possible to apply algorithms designed to reveal changes, events (for example, Karhunen-Loeve Transform [10], [15]) or to compute dedicated coefficients and indicators (correlation coefficients, determinants, eigenvalues – e.g. indexing of Web pages [9]). On the other hand, such datasets may contain redundant information like attributes which are highly correlated or have a small variance which may affects the reliability thus usefulness of the results. Considering large datasets containing legal texts, acts etc., dimensionality reduction seems to be vital step for the entire analysis.

Singular Value Decomposition method (SVD) is exploited in the area of dimensionality reduction, including natural language processing [5], [6] and Latent Semantic Analysis method [12], dedicated to acquiring semantic relations from large text datasets. The goal is to split the input m-by-n matrix A (see equation 2) into two orthogonal (unitary) matrices U (m-by-m) and V (n-by-n), and the diagonal one (S) of the same size as input matrix A, with nonnegative decreasing elements σ (singular values) [1].

$$A = USV^{\mathrm{T}}, \, U^{\mathrm{T}}U = I, \, V^{\mathrm{T}}V = I \tag{2}$$

Another approach to dimensionality reduction is associated with the use of the Principal Component Analysis method. Having a given n-by-n covariance matrix the principal component coefficients (loadings) are obtained. For further calculations the columns with the largest variances are taken [8].

In practice, the goal is to reduce the input matrix and to save as much information as possible. For this purpose, Latent Semantic Indexing algorithm (also called Latent Semantic Analysis, LSA) [2] may be utilized, in particular, to perform algebraic transformation of TDM matrix with SVD transformation (the most important LSA step [24]) to achieve lexical relations between terms.

3.2.3 Similarity between Objects vs. between Texts

The selection of a similarity method depends on the aim of analysis and the properties of the input dataset. Note, that inappropriate selection results in producing unreliable results.

An application of methods dedicated to assess the similarity between objects is based on the selection of the proper distance method or metric (function) among which the most popular is the Euclidean one. For the number of samples of each object p the Euclidean distance between x and y is defined as follows [7], [29]:

$$d_{E(y,x)} = \sqrt{\sum_{k=0}^{p}(y_k - x_k)^2 \frac{1}{p}} = \sqrt{\frac{(y-x)'(y-x)}{p}}, \tag{3}$$

The generalization of Euclidean, distance method is the Minkowski metric (also called m-norm):

$$d_{M(y,x)} = \left(\sum_{i=0}^{p}|y_i - x_i|^m / p\right)^{1/m} \tag{4}$$

Note that the same order of magnitude is assumed.

In the field of text/semantic analysis, the Jaccard coefficient and the cosine distance are exploited. The Jaccard coefficient for non-binary data, i.a. two vectors of attributes, is computed as $\frac{A \cap B}{A \cup B}$ (the size of the intersection of the dataset A and B related to the size of the union of the A and B) [7]. For the binary data, the simplest method of examining the similarity is the utilization of the factor $\frac{n(1,1)+n(0,0)}{n(1,1)+n(1,0)+n(0,1)+n(0,0)}$ which is the ratio of the variables for which the object has the same binary value to the total variables number p [7]. Also the Jaccard coefficient (see eq. 5) or the Dice one $\frac{2n(1,1)}{2n(1,1)+n(1,0)+n(0,1)}$ can be used.

$$S_J = \frac{n(1,1)}{n(1,1)+n(1,0)+n(0,1)} \tag{5}$$

Another approach is to use the cosine measure for text matching between two vectors (attributes) A and B (typically vectors: generated from documents (B) and a query one (A)), thus the angle between them (see eq. 6) [4].

$$S_C = \frac{A \cdot B}{\|A\| \cdot \|B\|} = \frac{\sum_{i=1}^{n} A_i B_i}{\sqrt{\sum_{i=1}^{n} A_i^2} \times \sqrt{\sum_{i=1}^{n} B_i^2}} \tag{6}$$

The high level of the similarity between objects (in this case, to maximum possible value is equal to 1) is indicated for a small angle between vectors.

The similarity analysis of the text datasets is performed for a list of terms retrieved/acquired from original document(s), represented as points generated with an algorithm. To be able to process a number of related documents input datasets are weighted in relation to the maximum term frequency and/or the number of analyzed documents [4], [17] (see eq. 1).

3.2.4   Semantic Similarity Analysis Based on WordNET

The development and availability of projects aimed at national lexical systems design, like WordNET [4],[18], entailed a possibility of similarity analyses of words through exploiting hierarchical relational structure between lexical units (lexemes). Note, that the coherence and completeness of such information is essential for legal text processing.

Several methods of the semantic similarity calculating based on Princeton WordNET network [3] have been proposed. In many cases specific implementation scenarios in the area of artificial intelligence research have been investigated [3], [14], [13]. One of the most efficient semantic similarity method, described in [13], is based on measuring of semantic distance between words within the WordNET network structure. This network consists of the sets of synonyms (synsets) which, similarly to lexical units, combines with each other by semantic relations: hyperonymy, meronymy, fuzzynymy etc.

A semantic word similarity analysis between legal texts may be supported with relations provided by WordNET. A similarity between lexemes is determined not only by the length of the path between them but also by the level of lexemes nesting.

For two lexemes: w_1 and w_2, semantic similarity is defined as follows [14]:

$$sim(w_1, w_2) = f(l, h) \tag{7}$$

where l denotes the shortest path between WordNET lexemes, and h denotes the nesting depth for the nearest superior lexeme and both the w_1 and w_2.

Formula (7) may be decomposed using the following functions (see eq. 8):

$$sim(w_1, w_2) = f_1(l) \cdot f_2(h) \tag{8}$$

where f_1 and f_2 denote functions for determining the length of the shortest path and the distance from the superior lexeme.

In this paper we consider relations which generate the similarity between two lexemes among WordNET network such as: synonymy, hyponymy, hyperonymy, meronymy, holonymy.

To calculate the similarity functions f_1 and f_2 the following formula (9) is used:

$$f_1(l) = \frac{1}{(1+\alpha l)^{6\alpha}}, f_2(h) = 1 - \frac{1}{(1+\beta h)^{4\beta}} \tag{9}$$

Therefore we obtain the following formula:

$$sim(w_1, w_2) = \frac{1}{(1+\alpha l)^{6\alpha}} \cdot \left(1 - \frac{1}{(1+\beta h)^{4\beta}}\right) \tag{10}$$

where α, β are the smoothing coefficients ($\alpha = 0.3$ and $\beta = 0.9$ are adequate to reach the best similarity between lexemes).

Figure 1 shows a diagram of the system dedicated to similarity analysis based on semantic relation among WordNET. Documents collection generates the input data for processing with morpho-syntactic tools and simultaneously extracting elements of structure from texts which are specific for the legal domain (preamble, paragraph, reference etc.). In the next step, obtained structures should be ordered to generate the term-by-document matrices. Note, that connection with WordNET network gave the opportunity to increasing the number of processed words. In addition, both the legal texts data aggregation (by the hyponymy relation) and meaning expansion (by the hyperonymy relation application) are possible. Input data are represented as categorized documents collections (document groups) which are similar to each other in the context of semantic relations.

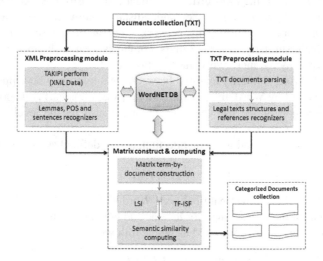

Fig. 1. A diagram of a system that performs the WordNET network connection

4   Sample Results

4.1   Overview

The goal of the part of research described in this section is to analyze a possibility of performing reliable similarity analysis of sample Polish legal texts based on approaches typically used for common language processing (in English).

For the prepared dataset of legal acts (see table 1) algebraic methods were applied to transform the original data into the set of terms suitable for further processing in order to similarity analysis. In this case, the Vector Space Model (VSM)[12] has been used to generate the incidence matrix of words in documents (term-document matrix, TDM).

---

[12] Salton G., Wong A., Yang C. S. (1975): A Vector Space Model for Automatic Indexing, Communications of the ACM, vol. 18, nr. 11, s. 613-620.

Table 1. Processed legal texts taken into analysis

Code/ No.	Document (legal text)	File size [MB]
D1	Ustawa o szczególnych warunkach sprzedaży konsumenckiej (An act on sell of consumers goods)[13]	0,66
D2	Ustawa o przeciwdziałaniu nieuczciwym praktykom rynkowym (An act against unfair market practices)[14]	1,50
D3	Ustawa o kredycie konsumenckim (An act on consumer credit)[15]	3,40
D4	Ustawa o ochronie konkurencji i konsumentów (An act on competition and consumer protection)[16]	4,8
D5	Ustawa o ubezpieczeniach obowiązkowych Ubezpieczeniowym Funduszu Gwarancyjnym i Polskim Biurze Ubezpieczycieli Komunikacyjnych (An act on Compulsory Insurance Insurance Guarantee Fund and Polish Motor Insurers)[17]	9,6
D6	Ustawa o działalności ubezpieczeniowej (Act on Insurance Business)[18]	14,1
D7	Ustawa o doradztwie podatkowym (Act on tax consulting)[19]	2,4
D8	Kodeks cywilny (Civil Code)[20]	17,5

Two sample similarity measures have been implemented: the Jaccard method and Cosine one (see Section 3.2.3). To this aim Java libraries (*org.apache.commons.math*) were used.

The input data (body of documents) was extracted from PDF files officially published. As a body of documents, a set of eight polish legal acts listed in Table 1 have been chosen.

All the collected legal texts have undergone a process of analysis and extraction of words, which consisted of the following stages:

a)   generation of morpho-syntactic text description from TXT format with *Takipi*[21], Polish texts tagger, to its XML output, read in the designed program;

b)   exploration and extraction of words specific to the legal texts using models based on the original text in TXT format (references to the "Dziennik Ustaw", referring to the date of enactment of legal documents, determining the names of offices and institutions etc.);

---

[13] http://isap.sejm.gov.pl/DetailsServlet?id=WDU20021411176
[14] http://isap.sejm.gov.pl/DetailsServlet?id=WDU20071711206
[15] http://isap.sejm.gov.pl/DetailsServlet?id=WDU20111260715
[16] http://isap.sejm.gov.pl/DetailsServlet?id=WDU20070500331
[17] http://isap.sejm.gov.pl/DetailsServlet?id=WDU20031241152
[18] http://isap.sejm.gov.pl/DetailsServlet?id=WDU20031241151
[19] http://isap.sejm.gov.pl/DetailsServlet?id=WDU19961020475
[20] http://isap.sejm.gov.pl/DetailsServlet?id=WDU19640160093
[21] Piasecki Maciej. Polish Tagger TaKIPI: Rule Based Construction and Optimisation. Task Quarterly, 2007, 11, 151-167. (http://nlp.pwr.wroc.pl/takipi/)

c) extracting tokens from the analysis (a) with the addition of specific terms
    of the legal texts and to create a final list of expressions (terms) used in
    the legal documents;
d) separating the expressions with stop-list for the Polish language and fre-
    quently occurring phrases specific to the legal texts (preamble, article,
    chapter, section, paragraph, reference etc.); text structures are ordered to
    create term-by-document matrices.

The input data are represented as categorized documents collections (document
groups) which are similar to each other according to semantic relations.

By establishing connection with WordNET network, based on semantic relations, the number of analyzed word was increased. In addition, both the terms
widening (by hiponimia relation) and focusing (by hiperonimia relation) became
possible.

The presented below experimental results are temporary (for the set of documents D1-D8 and for the given time instant) due to a large number of processed
words and WordNET database usage.

4.2 Experimental Results

Results obtained for the similarity analysis of sample legal texts for 2 types of
normalization (weighting TF/IDF and TDM normalization LSI) and 2 similarity
measures are shown in tables 1-4.

An application of TD/IDF normalization and Jaccard method indicates that the
most dissimilar document to others is D1 (probably due to the file size) while the
most similar is D6 (see table 2). The selection of cosine method instead of Jaccard
one produces different results (see table 3). In this case, the most similar is D4,
and the least one is D3. Note, that similar results (D3) are obtained with LSI normalization with both with Jaccard and cosine method (see tab.4-5).

Results generated with LSI normalization indicate higher similarity coefficient
values (see tables 4-5) related to TF/IDF normalization (tab. 2-3). For calculation
performed both with Jaccard (table 4) and cosine method (table 5) the results directly indicate D5 and D6 as the most similar documents.

Table 2. Results obtained with TF/IDF normalization and Jaccard method

	D1	D2	D3	D4	D5	D6	D7	D8
D1	1,0000	0,1218	0,1236	0,1196	0,1193	0,1085	0,0923	0,1361
D2	0,1218	1,0000	0,1346	0,1890	0,1225	0,1314	0,1087	0,1215
D3	0,1236	0,1346	1,0000	0,1413	0,1403	0,1361	0,1013	0,1310
D4	0,1196	0,1890	0,1413	1,0000	0,1828	0,2165	0,1992	0,1606
D5	0,1193	0,1225	0,1403	0,1828	1,0000	0,2757	0,1594	0,1772
D6	0,1085	0,1314	0,1361	0,2165	0,2757	1,0000	0,2043	0,1766
D7	0,0923	0,1087	0,1013	0,1992	0,1594	0,2043	1,0000	0,1290
D8	0,1361	0,1215	0,1310	0,1606	0,1772	0,1766	0,1290	1,0000
Sum:	0,8213	0,9295	0,9082	1,2089	1,1771	1,2490	0,9943	1,0320

Table 3. Results obtained with TF/IDF normalization and cosine method

	D1	D2	D3	D4	D5	D6	D7	D8
D1	1,0000	0,0491	0,0558	0,0768	0,0537	0,0447	0,0269	0,1532
D2	0,0491	1,0000	0,0563	0,2521	0,0544	0,0576	0,0387	0,0767
D3	0,0558	0,0563	1,0000	0,0735	0,0480	0,0463	0,0162	0,0649
D4	0,0768	0,2521	0,0735	1,0000	0,1113	0,1369	0,1066	0,1226
D5	0,0537	0,0544	0,0480	0,1113	1,0000	0,3759	0,0678	0,1526
D6	0,0447	0,0576	0,0463	0,1369	0,3759	1,0000	0,1000	0,1154
D7	0,0269	0,0387	0,0162	0,1066	0,0678	0,1000	1,0000	0,0572
D8	0,1532	0,0767	0,0649	0,1226	0,1526	0,1154	0,0572	1,0000
Sum:	0,4601	0,5849	0,3611	0,8797	0,8636	0,8766	0,4134	0,7425

Table 4. Results obtained with LSI normalization and Jaccard method. Similarity analysis between document D1-D8

	D1	D2	D3	D4	D5	D6	D7	D8
D1	1,0000	0,9735	0,4333	0,9026	0,7928	0,7747	0,7318	0,9439
D2	0,9735	1,0000	0,4464	0,8786	0,7714	0,7538	0,7128	0,9188
D3	0,4333	0,4464	1,0000	0,3858	0,3304	0,3219	0,3124	0,4062
D4	0,9026	0,8786	0,3858	1,0000	0,8790	0,8589	0,7998	0,9563
D5	0,7928	0,7714	0,3304	0,8790	1,0000	0,9761	0,8502	0,8404
D6	0,7747	0,7538	0,3219	0,8589	0,9761	1,0000	0,8655	0,8212
D7	0,7318	0,7128	0,3124	0,7998	0,8502	0,8655	1,0000	0,7742
D8	0,9439	0,9188	0,4062	0,9563	0,8404	0,8212	0,7742	1,0000
Sum:	5,5526	5,4552	2,6364	5,6610	5,4404	5,3722	5,0467	5,6609

Table 5. Results obtained with LSI normalization and cosine method

	D1	D2	D3	D4	D5	D6	D7	D8
D1	1,0000	0,9502	0,1195	0,9642	0,9946	0,9958	0,9975	0,9741
D2	0,9502	1,0000	0,4227	0,9988	0,9774	0,9748	0,9644	0,9960
D3	0,1195	0,4227	1,0000	0,3784	0,2216	0,2099	0,1725	0,3406
D4	0,9642	0,9988	0,3784	1,0000	0,9865	0,9844	0,9759	0,9992
D5	0,9946	0,9774	0,2216	0,9865	1,0000	0,9999	0,9976	0,9923
D6	0,9958	0,9748	0,2099	0,9844	0,9999	1,0000	0,9982	0,9907
D7	0,9975	0,9644	0,1725	0,9759	0,9976	0,9982	1,0000	0,9837
D8	0,9741	0,9960	0,3406	0,9992	0,9923	0,9907	0,9837	1,0000
Sum:	5,9959	6,2843	1,8652	6,2873	6,1700	6,1537	6,0898	6,2767

The results also show cross-correlations of document similarity in the set {D5, D6, D7, D8} (the Insurance Law and the Civil Code). The proposed algorithm has generated a group of legal acts from the similar semantic area (the Consumer Law; note, that the Civil Law has chapters related to the Insurance Law).

5   Conclusion and Future Work

Depending on the selection of algorithms employed for extracting terms from legal documents, dimensionality reduction and for preprocessing of input data, it can be assumed that there is a possibility of application numerical procedures (widely used in the area of common English language) to datasets containing Polish legal texts. The essential step towards obtaining reliable results is similarity method and the input matrix normalization numerical procedure selection.

In the paper, the preliminary work aimed at automatic polish legal texts processing was outlined. Taking into account presented considerations and sample results, further research will be aimed at polish legal acts clustering, in particular, through adjusting of numerical algorithms (briefly described above), and – additionally – performing analyses of semantic relations between terms, sentences and finally – legal acts.

References

1. Anderson, E., Bai, Z., Bischof, C., Blackford, S., Demmel, J., Dongarra, J., Croz, J., Du, G.A., Hammarling, S., McKenney, A., Sorensen, D.: LAPACK User's Guide, 3rd edn. SIAM, Philadelphia (1999),
   http://www.netlib.org/lapack/lug/lapack_lug.html
2. Broda, B., Piasecki, M.: SuperMatrix: a General Tool for Lexical Semantic Knowledge Acquisition. In: Speech and Language Technology, pp. 239–254. Polish Phonetics Assocation (2008)
3. Budanitsky, A., Hirst, G.: Semantic Distance in WordNet: An Experimental, Application-Oriented Evaluation of Five Measures. In: Proc. of the Workshop WordNet and Other Lexical Resources, Second Meeting of the North Am. Chapter of the Association for Computational Linguistics (2001)
4. Charikar, M.S.: Similarity Estimation Techniques from Rounding Algorithms. In: Reif, J. (ed.) STOC 2002 Proceedings of the Thirty-fourth Annual ACM Symposium on Theory of Computing. ACM (2002)
5. Deerwester, S.C., Dumais, S.T., Landauer, T.K., Furnas, G.W., Harshman, R.A.: Indexing by latent semantic analysis. Journal of the American Society of Information Science 41(6), 391–407 (1990)
6. Deerwester, S., et al.: Improving Information Retrieval with Latent Semantic Indexing. In: Proc. of the 51st Annual Meeting of the American Society for Information Science, vol. 25, pp. 36–40 (1988)
7. Hand, D., Mannila, H., Smyth, P.: Principles of Data Mining. MIT Press (2001)
8. Jolliffe, I.T.: Principal Component Analysis, 2nd edn. Springer, New York (2002)

9. Kamvar, S.D., Haveliwala, T.H., Manning, C.D., Golub, G.H.: Extrapolation methods for accelerating PageRank computations. In: Proc. of the WWW 2003 Proc. of the 12th International Conference on World Wide Web, pp. 261–270. ACM, New York (2003)
10. Karhunen, K.: Zur Spektraltheorie Stochastischer Prozesse. Annales Academiae Scientiarum Fennicae, Series A1, Mathematica-Physica 34, 1–7 (1946)
11. Kisz, A.: Model cybernetyczny powstawania i działania prawa, Wrocław (1970)
12. Landauer, T., Dumais, S.: A solution to Plato's problem: The latent semantic analysis theory of acquisition. Psychological Review 104(2), 211–240 (1997)
13. Li, Y.H., Bandar, Z., McLean, D.: An Approach for Measuring Semantic Similarity Using Multiple Information Sources. IEEE Trans. Knowledge and Data Eng. 15(4), 871–882 (2003)
14. Li, Y., McLean, D., Bandar, Z.A., O'Shea, J.D., Crockett, K.: Sentence Similarity Based on Semantic Nets and Corpus Statistics. IEEE Transaction of Knowledge and Data Engineering 18(8), 1138–1150 (2006)
15. Loève, M.M.: Probability Theory. VanNostrand, Princeton (1955)
16. Malinowski, A.: Wstęp do badań cybernetycznych w prawoznawstwie, Warszawa (1977)
17. Markines, B., Cattuto, C., Menczer, F., Benz, D., Hotho, A., Stumme, G.: Evaluating Similarity Measures for Emergent Semantics of Social Tagging. In: WWW 2009 Proceedings of the 18th International Conference on World Wide Web. ACM (2009)
18. Maziarz, M., Piasecki, M., Szpakowicz, S.: Approaching plWordNet 2.0. In: Proc. of the 6th Global Wordnet Conference, Matsue, Japan, January 9-13 (2012) (accepted for publishing)
19. Palmirani, M., Cervone, L., Vitali, F.: A Legal Document Ontology: The Missing Layer in Legal Document Modeling. In: Sartor, G. (ed.) Approaches to Legal Ontologies. Thwoeiwa, Domains, Methodologies. Springer, Dordrecht (2011)
20. Potiopa, P.: Methods and tools for the automatic processing of textual information and its use in the process of knowledge management. Automatyka 15(2), 409–419 (2011)
21. Salton, G., McGill, M.: Introduction to Modern Information Retrieval. McGraw-Hill (1983)
22. Sartor, G.: Introduction: ICT and Legislation in the Knowledge Society. In: Sartor, G., Palmirani, M., Francesconi, E., Biasiotti, M.A. (eds.) Legislative XML for Semantic Web, p. 23. Springer, Dordrecht (2012)
23. Sobczak, K.: Prawo a informatyka. Wydawnictwo Prawnicze, Warszawa (1978)
24. Stewart, G.W.: On the Early History of the Singular Value Decomposition. SIAM Review 35(4), 551–566 (1993)
25. Studnicki, F.: Cybernetyka a prawo, Warszawa (1969)
26. Turtle, H.: Text Retrieval in the Legal World. Artificial Intelligence and Law (3) (1995)
27. Vitali, F.: A Standard-Based Approach for the Management of Legislative Documents. In: Legislative XML for the Semantic Web, Law, Governance and Technology Series, vol. 4, p. 44 (2011)
28. Wiewiórowski, W.R., Wierczyński, G.: Informatyka prawnicza. Technologia informacyjna dla prawników i administracji publicznej. Oficyna – Wolters Kluwer business, Warszawa (2008)
29. Yang, K., Shahabi, C.: A PCA-Based Similarity Measure for Multivariate Time Series. In: ACM International Workshop on Multimedia Databases (2004)

AI Approach to Formal Analysis of BPMN Models: Towards a Logical Model for BPMN Diagrams

Antoni Ligęza and Tomasz Potempa

Abstract. Modeling Business Processes has become a challenging issue of today's Knowledge Management. As such it is a core activity of Knowledge Engineering. There are two principal approaches to modeling such processes, namely Business Process Modeling and Notation (BPMN) and Business Rules (BR). Both these approaches are to certain degree complementary, but BPMN seems to become a standard supported by OMG. In this paper we investigate how to build a logical model of BPMN using logic, logic programming and rules. The main focus in on logical reconstruction of BPMN semantics which is necessary to define some formal requirements on model correctness enabling formal verification of such models.

1 Introduction

Knowledge has become a valuable and critical resource of contemporary organizations. In fact, possession of valid, most complete, up-to-date and essential knowledge has become a decisive factor of success in the so competitive market.

Unfortunately — or no — human possession and processing of knowledge turns out to be fairly inefficient for a number of reasons. Some most obvious ones include difficulties with knowledge sharing, storage, and efficient execution. Hence, *Knowledge Management* (KM) must be supported with tools

Antoni Ligęza
AGH University of Science and Technology,
al. A. Mickiewicza 30, 30-059 Krakow, Poland
e-mail: ligeza@agh.edu.pl

Tomasz Potempa
Higher School of Tarnów
ul. Mickiewicza 8, 33-100 Tarnów, Poland
e-mail: tpotempa@gmail.com

M. Mach-Król and T. Pełech-Pilichowski (eds.), *Advances in Business ICT*,
Advances in Intelligent Systems and Computing 257,
DOI: 10.1007/978-3-319-03677-9_5, © Springer International Publishing Switzerland 2014

and techniques coming from Software Engineering and Knowledge Engineering universe.

Design, development and analysis of progressively more and more complex business processes require advanced methods and tools. Two generic modern approaches to modeling such processes have recently gained wider popularity. These are the *Business Process Modeling and Notation* [29, 32, 1], or BPMN for short, and *Business Rules* [28, 2, 7]. Although aimed at a common target, both of the approaches are rather mutually complementary and offer somewhat distinctive features enabling process modeling [9] and executing [20].

BPMN as such constitutes a set of graphical symbols, such as links modeling workflow, various splits and joins, events and boxes representing data processing activities. It forms a transparent visual tool for modeling complex processes promoted by OMG [25]. What is worth underlying is the expressive power of current BPMN. In fact it allows for modeling conditional operations, loops, event-triggered actions, splits and joins of data flow paths and communication processes [27]. Moreover, modeling can take into account several levels of abstraction which enables hierarchical approach.

An important issue about BPMN is that it covers three important aspects of any business process; these are:

- *data processing* or *data flow* specification; this includes input, output and internal data processing,

- *inference control* or *workflow control*; this includes diagrammatic specification of the process with partial ordering, switching and merging of flow,

- *structural representation* of the process as a whole; this allows for a visual representation at several levels of hierarchy.

BPMN as such can be considered as *procedural knowledge representation*; a BPMN diagram represents in fact a set of interconnected procedures. On the other hand, the workflow diagram, however, although it provides transparent, visual picture of the process, due to lack of formal model semantics makes attempts at more rigorous analysis problematic. Further, even relatively simple inference requires a lot of space for representation; there is no easy way to specify declarative knowledge, e.g. in the form of rules.

Business Rules (BR), also promoted by OMG [23, 24], offer an approach to specification of knowledge in a *declarative* manner. The way the rules are applied is left over to the user when it comes to rule execution. Hence, rules can be considered as *declarative knowledge specification*; inference control is normally not covered by basic rules.

Note that rules can fulfill different roles in the system [23]. Some three most important ones cover:

- *declarative knowledge specification* for inference of new facts,

- *integrity constraints* for preserving consistency of the knowledge base, and

- *meta-rules* i.e. rules defining how to use other rules; this may include partial *control knowledge specification* for improving efficiency of the inference process.

Some modern classifications cover the following types of rules:

- facts – rules defining true statement (no conditional part),
- definition rules – for defining terms and notions in use,
- integrity rules – rules defining integrity constraints,
- production rules – for derivation of new facts,
- reaction rules – rules triggered by events, reactive rules or ECA rules,
- transformation rules – rules defining possible transformations, term-rewriting rules; they may include numerical recipe rules,
- data processing rules – rules defining how particular data are to be transformed; these include numerical processing rules,
- control rules – in fact meta rules used for inference process control,
- meta rules – other rules defining how to use basic rules.

Rules, especially when grouped into decision modules (such as decision tables) are easier to analyze, however, the possibility of analysis depends on the accepted *knowledge representation language*, and in fact – the logic in use [14]. Formal models of rule-based systems and analysis issues are discussed in detail in [14].

Note that the two approaches are to certain degree complementary: Business Rules provide declarative specification of domain knowledge, which can be encoded into a BPMN model. On the other hand, a careful analysis of a BPMN diagram allows to extract certain rules governing the business. However, there is no consistent study on that possibility of mutual conversion. The main problems seem to consist in lack of formal specification of KR language. A some under-identification of BPMN.

The main common problem of BPMN is the lack of a *formal declarative model* defining precisely the semantics and logic behind the diagram. Hence defining and analyzing correctness of BPMN diagrams (e.g. in terms of termination or determinism) is a hard task. There are only few papers undertaking the issues of analysis and verification of BPMN diagrams [5, 26, 27]. However, the analysis is performed mostly at the *structural* level and does not take into account the semantics of dataflow and control knowledge.

In this paper we follow the ideas initially presented in [15] and extend the work described in [16]. An attempt at defining foundations for a more formal, logical, declarative model of the most crucial elements of BPMN diagrams is undertaken. We pass from logical analysis of BPMN components to their logical models, properties and representation in PROLOG. Some prototyped procedures for checking correct data flow are also presented. The model is aimed at enabling definition and further analysis of selected formal properties of a class of restricted BPMN diagrams. The analysis should take into account properties constituting reasonable criteria of correctness. The focus is on development of a formal, declarative model of BPMN components and its overall structure. In fact, a combination of recent approaches to development and verification of rule-based systems [21, 17, 19] seems to have potential influence on BPMN analysis.

2 A Basic Structural Model for BPMN

In this section a simplified structural model of BPMN diagrams is put forward. It constitutes a restricted abstraction of crucial intrinsic workflow components. As for events, only start and termination events are taken into account. Main knowledge processing units are activities (or tasks). Workflow control is modeled by subtypes of gateways: split and join operations. Finally, workflow sequence is modeled by directed links. No time/temporal aspect are considered. The following elements will be taken into consideration:

- $\mathbb{S}$ — a non-empty set of *start events* (possibly composed of a single element),
- $\mathbb{E}$ — a non-empty set of *end events* (possibly composed of a single element),
- $\mathbb{T}$ — a set of *activities* (or *tasks*); a task $T \in \mathbb{T}$ is a finite process with single input and single output, to be executed within a finite interval of time,
- $\mathbb{G}$ — a set of *split gateways* or *splits*, where branching of the workflow takes place; three disjoint subtypes of splits are considered:
    - $\mathbb{GX}$ — a set of *exclusive splits* where one and only one alternative paths can be followed (a split of EX-OR type),
    - $\mathbb{GP}$ — a set of *parallel splits* where all the paths of the workflow are to be followed (a split of *AND* type or a *fork*), and
    - $\mathbb{GO}$ — a set of *inclusive splits* where one or more paths should be followed (a split of *OR* type).
- $\mathbb{M}$ — a set of *merge gateways* or *joins* node of the diagram, where two or more paths meet; three further disjoint subtypes of merge (join) nodes are considered:
    - $\mathbb{MX}$ — a set of *exclusive merge* nodes where one and only one input path is taken into account (a merge of EX-OR type),
    - $\mathbb{MP}$ — a set of *parallel merge* nodes where all the paths are combined together (a merge of *AND* type), and
    - $\mathbb{MO}$ — a set of *inclusive merge* nodes where one or more paths influence the subsequent item (a merge of *OR* type).
- $\mathbb{F}$ — a set of workflow links, $\mathbb{F} \subseteq \mathbb{O} \times \mathbb{O}$, where $\mathbb{O} = \mathbb{S} \cup \mathbb{E} \cup \mathbb{T} \cup \mathbb{G} \cup \mathbb{M}$ is the join set of objects. All the component sets are pairwise disjoint.

The splits and joins depend on logical conditions assigned to particular branches. It is assumed that there is defined a partial function $\texttt{Cond}: \mathbb{F} \to \mathbb{C}$ assigning logical formulae to links. In particular, the function is defined for links belonging to $\mathbb{G} \times \mathbb{O} \cup \mathbb{O} \times \mathbb{M}$, i.e. outgoing links of split nodes and incoming links of merge nodes. The conditions are responsible for workflow control. For intuition, a simple BPMN diagram is presented in Fig. 1.

In order to ensure *correct structure* of BPMN diagrams a set of restrictions on the overall diagram structure is typically defined; they determine the so-called *well-formed diagram* [27].

Note however, that a well-formed diagram does not assure that for any input knowledge the process can be executed leading to a (unique) solution. This further depends on the particular input data, its transformation during processing, correct work of particular objects, and correct control defined by the branching/merging conditions assigned to links.

3 An Example of BPMN Diagram with Rules

In order to provide intuitions, the theoretical considerations will be illustrated with a simple example process. The process goal is to establish the so-called *set-point* temperature for a thermostat system [22]. The selection of the particular value depends on the season, whether it is a working day or not, and the time of the day.

Consider the following set of declarative rules specifying the process. There are eighteen inference rules (production rules):

Rule 1: $aDD \in \{monday, tuesday, wednesday, thursday, friday\} \longrightarrow aTD = wd.$
Rule 2: $aDD \in \{saturday, sunday\} \longrightarrow aTD = wk.$
Rule 3: $aTD = wd \wedge aTM \in (9, 17) \longrightarrow aOP = dbh.$
Rule 4: $aTD = wd \wedge aTM \in (0, 8) \longrightarrow aOP = ndbh.$
Rule 5: $aTD = wd \wedge aTM \in (18, 24) \longrightarrow aOP = ndbh.$
Rule 6: $aTD = wk \longrightarrow aOP = ndbh.$
Rule 7: $aMO \in \{january, february, december\} \longrightarrow aSE = sum.$
Rule 8: $aMO \in \{march, april, may\} \longrightarrow aSE = aut.$
Rule 9: $aMO \in \{june, july, august\} \longrightarrow aSE = win.$
Rule 10: $aMO \in \{september, october, november\} \longrightarrow aSE = spr.$
Rule 11: $aSE = spr \wedge aOP = dbh \longrightarrow aTHS = 20.$
Rule 12: $aSE = spr \wedge aOP = ndbh \longrightarrow aTHS = 15.$
Rule 13: $aSE = sum \wedge aOP = dbh \longrightarrow aTHS = 24.$
Rule 14: $aSE = sum \wedge aOP = ndbh \longrightarrow aTHS = 17.$
Rule 15: $aSE = aut \wedge aOP = dbh \longrightarrow aTHS = 20.$
Rule 16: $aSE = aut \wedge aOP = ndbh \longrightarrow aTHS = 16.$
Rule 17: $aSE = win \wedge aOP = dbh \longrightarrow aTHS = 18.$
Rule 18: $aSE = win \wedge aOP = ndbh \longrightarrow aTHS = 14.$

Let us briefly explain these rules. The first two rules define if we have today (aTD) a workday (wd) or a weekend day (wk). Rules 3-6 define if the operation hours (aOP) are during business hours (dbh) or not during business hours ($ndbh$); they take into account the workday/weekend condition and the current time (hour). Rules 7-10 define the season (aSE) is summer (sum), autumn (aut), winter (win) or spring (spr). Finally, rules 11-18 define the

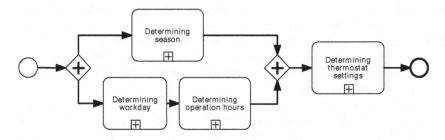

Fig. 1 An example BPMN diagram — top-level specification of the thermostat
system

precise setting of the thermostat ($aTHS$). Observe that the set of rules is flat;
basically no control knowledge is provided.

Now, let us attempt to visualize a business process defined with these rules.
A BPMN diagram of the process is specified in Fig. 1.

After start, the process is split into two independent paths of activities.
The upper path is aimed at determining the current season[1] (aSE; it can take
one of the values {sum, aut, win, spr}; the detailed specification is provided
with rules 7-10 below). A more visual specification of this activity with an
appropriate set of rules is shown in Fig. 2.

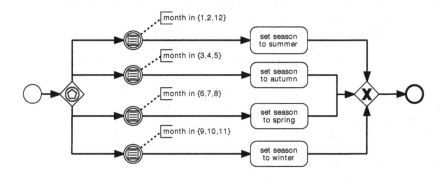

Fig. 2 An example BPMN diagram — detailed specification a BPMN task

The lower path activities determine whether the day (aDD) is a workday
($aTD = wd$) or a weekend day ($aTD = wk$), both specifying the value of
today (aTD; specification provided with rules 1 and 2), and then, taking
into account the current time (aTM), whether the operation (aOP) is during business hours ($aOP = dbh$) or not ($aOP = ndbh$); the specification is
provided with rules 3-6. This is illustrated with Fig. 3 and Fig. 4.

---

[1] For technical reasons all attribute names used in this example start with lower-
case 'a'.

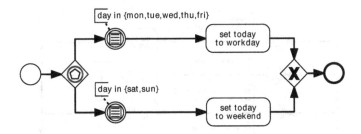

Fig. 3 An example BPMN diagram — detailed specification of determining the day task

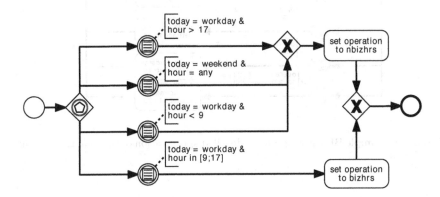

Fig. 4 An example BPMN diagram — detailed specification of working hours task

Finally, the results are merged together and the final activity consists in determining the thermostat settings (*aTHS*) for particular season (*aSE*) and time (*aTM*) (the specification is provided with rules 11-18). This is illustrated with Fig. 5.

Even in this simple example, answers to the following important questions are not obvious:

1. *data flow correctness*: is any of the four tasks/activities specified in a correct way? Will each task end with producing the desired output for any admissible input data?
2. *split consistency*: will the workflow possibly explore all the paths after a split? Will it always explore at least one?
3. *merge consistency*: will it be always possible to merge knowledge coming from different sources at the merge node?
4. *termination/completeness*: does the specification assure that the system will always terminate producing some temperature specification for *any* admissible input data?
5. *determinism*: will the output setting be determined in a unique way?

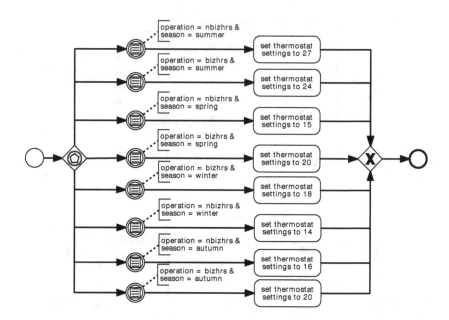

Fig. 5 An example BPMN diagram — detailed specification of the final thermostat setting task

Note that we do not ask about *correctness* of the result; in fact, the rules, embedded into a BPMN diagram, provide a kind of *executable specification*, so there is no reference point to claim that the final output is correct or not.

4 Logical Analysis of BPMN Diagrams

A BPMN diagram can model quite complex processes. Apart from *external consistency* validation (i.e. whether or not the diagram models correctly the external system in a *complete* way, does not introduce any non-existent features, and there is an *isomorphism* between those two), an important issue is the *internal consistency* requirement for *correct structure* of the diagram and *correct workflow specification*. The first one refers to the static specification of components and their connections. The second one consists in correct work of the structure for all admissible input data specification.

The structural correctness is defined by requirements for *well-formed BPMN diagram*. However, even having correct structure, the process can go wrong due to unserved data or wrong workflow control, for example. Below, an attempt is made at specification of some minimal requirements for (i) correct work of process components (tasks), (ii) assuring data flow, (iii) correct work of splits, (iv) correct work of merge nodes, and finally — (v) termination of the overall process.

A more complex issue consists in verification of internal workflow correctness. The analysis must take into account, at least the following aspects:

1. Local correctness requirements:

   a. Correct specification and work of process components performing activities,
   b. Correct specification of data flow,
   c. Correct specification and work of splits,
   d. Correct specification and work of joins.

2. Global correctness requirements:

   a. No deadlocks — the system must work for all admissible input data,
   b. Termination — the system must terminate work within a finite period of time,
   c. Determinism — the results should be repeatable for repeated input data (ambiguous rules, hazards, races).

Further requirements may refer to features such as *minimal representation, optimal decomposition, robustness,* and *optimal results* and *optimal execution.* In order to answer these questions one must (i) assure the correct work of all the components (activities/processes), (ii) assure the correctness of data flow between components, (iii) assure correct inference control (w.r.t. split and join operations), (iv) check if the static structure of the diagram is correct and, finally (v) check if all this combined together will work, i.e. the inference process is not blocked at some point (e.g. due to a deadlock).

4.1 Component Correctness

In this section we put forward some minimal requirements defining correct work of rule-based process components performing BPMN activities. Each such component is composed of a set of inference rules, designed to work within a the same context; in fact, preconditions of the rules incorporate the same attributes. In our example we have four such components: determining workday (rules 1-2), determining operation hours (rules 3-6), determining season (rules 7-10) and determining the thermostat setting (rules 11-18).

In general, the outermost logical model of a component T performing some activity/task can be defined as a triple of the form:

$$T = (\psi_T, \varphi_T, \mathcal{A}), \tag{1}$$

where ψ_T is a formula defining the restrictions on the component input, φ_T defines the restrictions for component output, and $\mathcal{A}$ is an algorithm which for a given input satisfying ψ_T produces an (desirably uniquely defined) output,

satisfying φ_T. For intuition, ψ_T and φ_T define a kind of a 'logical tube' —
for every input data satisfying ψ_T (located at the entry of the tube), the
component will produce and output satisfying φ_T (still located within the
tube at its output). The precise recipe for data processing is given by an
algorithm $\mathcal{A}$.

The specification of a rule-based process component given by (1) is considered *correct*, if and only if for any input data satisfying ψ_T the algorithm $\mathcal{A}$
produces an output satisfying φ_T. It is further *deterministic* (unambiguous)
if the generated output is unique for any admissible input.

For example, consider the component determining operation hours. Its
input restriction formula ψ_T is the disjunction of precondition formulae $\psi_3 \lor \psi_4 \lor \psi_5 \lor \psi_6$, where ψ_i is a precondition formula for rule i. We have $\psi_T = ((aTD = wd) \land (aTM \in [0,8] \lor aTM \in [9,17] \lor aTM \in [18,24])) \lor (aTD = wk)$. The output restriction formula is given by $\varphi_T = (aOP = dbh) \lor (aOP = ndbh)$. The algorithm is specified directly by the rules; rules are in fact a kind
of *executable specification*.

In order to be sure that the produced output is unique, the following
mutual exclusion condition should hold:

$$\not\models \psi_i \land \psi_j \tag{2}$$

for any $i \neq j$, $i, j \in \{1, 2, \ldots, k\}$. A simple analysis shows that the four rules
have mutually exclusive preconditions, and the joint precondition formula ψ_T
covers any admissible combination of input parameters; in fact, the subset of
rules is locally *complete* and *deterministic* [14].

4.2   Correct Flow of Data

In our example we consider only rule-based components. Let ϕ define the
context of operation, i.e. a formula defining some restrictions over the current state of the knowledge-base that must be satisfied before the rules of a
component are explored. For example, ϕ may be given by $\varphi_{T'}$ of a component T' directly preceding the current one. Further, let there be k rules in
the current component, and let ψ_i denote the joint precondition formula (a
conjunction of atoms) of rule i, $i = 1, 2, \ldots, k$. In order to be sure that at
least one of the rules will be fired, the following condition must hold:

$$\phi \models \psi_T, \tag{3}$$

where $\psi_T = \psi_1 \lor \psi_2 \lor \ldots \lor \psi_k$ is the disjunction of all precondition formulae
of the component rules. The above restriction will be called the *funnel principle*. For intuition, if the current knowledge specification satisfies restriction
defined by ϕ, then at least one of the formula preconditions must be satisfied
as well.

For example, consider the connection between the component determining workday and the following it component determining operation hours. After leaving the former one, we have that $aTD = wd \vee aTD = wk$. Assuming that the time can always be read as an input value, we have $\phi = (aTD = wd \vee aTD = wk) \wedge aTM \in [0, 24]$. On the other hand, the disjunction of precondition formulae $\psi_3 \vee \psi_4 \vee \psi_5 \vee \psi_6$ is given by $\psi_T = (aTD = wd) \wedge (aTM \in [0, 8] \vee aTM \in [9, 17] \vee aTM \in [18, 24])) \vee aTD = wk$. Obviously, the funnel condition given by (3) holds.

4.3 Correct Splits

An exclusive split $GX(q_1, q_2, \ldots q_k) \in \mathbb{GX}$ with k outgoing links is modeled by a fork structure assigned excluding alternative of the form:

$$q_1 \veebar q_2 \veebar \ldots \veebar q_k,$$

where $q_i \wedge q_j$ is always false for $i \neq j$. An exclusive split can be considered correct if and only if at least one of the alternative conditions is satisfied. We have the following logical requirement:

$$\models q_1 \vee q_2 \vee \ldots \vee q_k, \tag{4}$$

i.e. the disjunction is in fact a tautology. In practice, to assure (4), a pre-defined exclusive set of conditions is completed with a default q_0 condition defined as $q_0 = \neg q_1 \wedge \neg q_2 \wedge \ldots \wedge \neg q_k$; obviously, the formula $q_0 \vee q_1 \vee q_2 \vee \ldots \vee q_k$ is a tautology.

Note that in case when an input restriction formula ϕ is specified, the above requirement given by (4) can be relaxed to:

$$\phi \models q_1 \vee q_2 \vee \ldots \vee q_k. \tag{5}$$

An inclusive split $GO(q_1, q_2, \ldots q_k) \in \mathbb{GO}$ is modeled as disjunction of the form:

$$q_1 \vee q_2 \vee \ldots \vee q_k,$$

An inclusive split to be considered correct must also satisfy formula (4), or at least (5). As before, this can be achieved through completing it with the q_0 default formula.

A parallel split $GP(q_1, q_2, \ldots q_k) \in \mathbb{GP}$ is referring to a fork-like structure, where all the outgoing links should be followed in any case. For simplicity, a parallel split can be considered as an inclusive one, where all the conditions assigned to outgoing links are set to *true*.

Note that, if ϕ is the restriction formula valid for data at the input of the split, then any of the output restriction formula is defined as $\phi \wedge q_i$ for any of the outgoing link i, $i = 1, 2, \ldots, k$.

4.4 Correct Merge

Consider a workflow merge node, where k knowledge inputs satisfying restrictions $\phi_1, \phi_2, \ldots, \phi_k$ respectively meet together, while the selection of particular input is conditioned by formulae $p_1, p_2, \ldots, p_k$, respectively.

An exclusive merge $MX(p_1, p_2, \ldots, p_k) \in \mathbb{MX}$ of k inputs is considered correct if and only if the conditions are pairwise disjoint, i.e.

$$\not\models p_i \wedge p_j \tag{6}$$

for any $i \neq j$, $i, j \in \{1, 2, \ldots, k\}$. Moreover, to assure that the merge works, at least one of the conditions should hold:

$$\models p_1 \vee p_2 \vee \ldots \vee p_k, \tag{7}$$

i.e. the disjunction is in fact a tautology. If the input restrictions $\phi_1, \phi_2, \ldots, \phi_k$ are known, condition (7) might possibly be replaced by $\models (p_1 \wedge \phi_1) \vee (p_2 \wedge \phi_2) \vee \ldots \vee (p_k \wedge \phi_k)$.

Note that in case a join input restriction formula ϕ is specified, the above requirement can be relaxed to:

$$\phi \models p_1 \vee p_2 \vee \ldots \vee p_k, \tag{8}$$

and if the input restrictions $\phi_1, \phi_2, \ldots, \phi_k$ are known, it should be replaced by $\phi \models (p_1 \wedge \phi_1) \vee (p_2 \wedge \phi_2) \vee \ldots \vee (p_k \wedge \phi_k)$.

An inclusive merge $MO(p_1, p_2, \ldots, p_k) \in \mathbb{MO}$ of k inputs is considered correct if one is assured that the merge works — condition (7) or (8) hold.

A parallel merge $MP \in \mathbb{MP}$ of k inputs is considered correct by default. However, if the input restrictions $\phi_1, \phi_2, \ldots, \phi_k$ are known, a consistency requirement for the combined out takes the form that ϕ must be consistent (satisfiable), where:

$$\phi = \phi_1 \wedge \phi_2 \wedge \ldots \wedge \phi_k \tag{9}$$

An analogous requirement can be put forward for the active links of an inclusive merge.

$$\models p_1 \wedge p_2 \wedge \ldots \wedge p_k, \tag{10}$$

i.e. the conjunction is in fact a tautology, or at least

$$\phi \models p_1 \wedge p_2 \wedge \ldots \wedge p_k. \tag{11}$$

In general, parallel merge can be made correct in a trivial way by putting $p_1 = p_2 = \ldots = p_k = true$.

Note that even correct merge leading to a satisfiable formula assure only passing the merge node; the funnel principle must further be satisfied with respect to the following-in-line object. To illustrate that consider the input

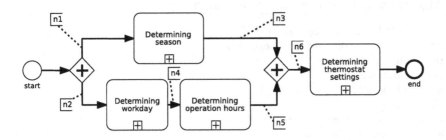

Fig. 6 An example BPMN diagram — top-level specification of thermostat system with the additional Prolog annotations

of the component determining thermostat setting. This is the case of parallel merge of two inputs. The joint formula defining the restrictions on combined output of the components for determining season and determining operation hours is of the form:

$$\phi = (aSE = sum \lor aSE = aut \lor aSE = win \lor$$
$$aSE = spr) \land (aOP = dbh \lor aOP = ndbh).$$

A simple check of all possible combinations of season and operation hours shows that all the eight possibilities are covered by preconditions of rules 11-18; hence the funnel condition (3) holds.

5 BPMN Model in Prolog

In this section a Prolog model for the example BPMN diagram is presented. It enables logical analysis of the diagram. Below, samples of Prolog code are listed.

A BPMN diagram is represented as a complex graph with different types of nodes. Each specific component of BPMN is mapped into a Prolog fact. In order to model the structure of a BPMN diagram in a declarative way it is proposed to use a set of PROLOG facts. A generic form can be as follows:

```
component_type(<id>,
    <input_node>,<input-formula>,
    <output_node>,<output-formula>).
```

In practice, such facts can be of different structure for different components.

The Thermostat example presented in Fig. 6 is defined with the Prolog code in Listing 1.

Algorithm 1. BPMN example coding in Prolog

```
%%%% BPMN Knowledge Base %%%%
%%% Start & end nodes declaration
init_node(start).
end_node(end).
nodes([n1,n2,n3,n4,n5,n6]).
%%% Tasks
% task(<id>,<input_node>,
%       <input_formula>,
%       <output_node>,
%       <output_formula>).
task(ds,n1
      [[mspr],[msum],[maut],[mwin]],
      n3,[[spr],[sum],[aut],[win]]).
task(dw,n2,[[d15],[d67]],n4,
      [[twr],[twe]]).
task(do,n4,[[twr,t18],[twr,t8],
      [twr,t917],[twe]],n5,
      [[nonbiz],[biz]]).
task(dt,n6,[[nonbiz,spr],[nonbiz,sum],
      [nonbiz,aut],[nonbiz,win],
      [biz,spr],[biz,sum],[biz,aut],
      [biz,win]],end,
      [[t14],[t15],[t16],[t18],[t20],
      [t24],[t27]]).
% split2(<id>,<split_type>,
%        <input_node>,
%        (<output_node>,
%         <logical_condition>),
%        (<output_node>,
%         <logical_condition>)).
split2(s1,and,start,(n1,true),
      (n2,true)).
% merge2(<id>,<merge_type>,
%        <input_node>,
%        <input_node>,
%        <output_node>).
merge2(m1,and,n3,n5,n6).
```

There are four tasks defined:

ds *define season* with input node $n1$, output node $n3$, input formula being a
DNF - disjunction of all four precondition formulas of the four rules, and
output formula being DNF - a disjunction of the conclusions of the rules,

dw*define workday* with input node $n2$, output node $n4$, input formula being a DNF - disjunction of all two precondition formulas of the two rules, and output formula being DNF - a disjunction of the conclusions of the rules,

do *define operation* with input node $n4$, output node $n5$, input formula being a DNF - disjunction of all four precondition formulas of the four rules, and output formula being DNF - a disjunction of the conclusions of the rules,

dt *define temperature* with input node $n1$, output node $n3$, input formula being a DNF - disjunction of all eight precondition formulas of the eight rules, and output formula being DNF - a disjunction of the conclusions of the rules.

There are also one split node, from start to $n1$ and $n2$, and one merge node, from $n3$ and $n5$ to $n6$.

As an example of analysis let us present a code for verification of data flow (the funnel condition) between tasks *dw* and *do*; the intuition behind is that any output generated by *dw* should be accepted and further processed by *do*. The sample code is presented in Listing 2.

After calling the funnel check for the internal node $n4$ the system proves that the output formula of task *dw* implies the input formula of task *do*. The output of the program, confirming the result of the check is given below.

Other static checks for data flow have been implemented and tested as well.

6 Related Works

To the best of our knowledge, there are no related works which define BPMN model in the Prolog language. The only one approach that uses Prolog by Androček [4] significantly differs from our work. Androček used Prolog for specification of business processes, however, he used it to assess the cost and time of running a process and to identify potential problems with resources. Thus, he defined the model for simulation purposes.

Similar attempts to formalization of BPMN models were carried out by Lam [12], Andersson et al. [3], Wong and Gibbons [33] as well as Dijkman and Van Gorp [6]. Other attempts to formalization of process models, however not concerning BPMN, were carried out by Gruhn and Laue [8].

The BPMN model defined by Lam in [12] was used to check the diagrams against the properties specified by a user for a particular model [13]. In our case, the properties, which are correctness requirements, can be used for any BPMN model.

In the case of declarative model presented in [3], they introduced the notion of activity dependency model, which identifies, classifies, and relates activities needed for executing and coordinating value transfers. In particular, in their model, relations between activities can be specified in terms of notions like resource flow, trust, coordination, and reciprocity. The four types of dependencies which can be identified (flow, trust, trigger, and duality dependencies) are

---

Algorithm 2. BPMN example coding in Prolog: funnel between tasks

---

```
%%% Funnel condition checking for
%%% nodes among tasks
funnel(N) :-
    task(IDOUT,_,_,N,FOUT),
    task(IDIN,N,FIN,_,_),
    implies(FOUT,FIN),
    write(IDOUT),write('—>'),write(N),

    write('—>'),write(IDIN),nl.

%%% Definition of implication
%%% for two DNF
implies(true,_) :- !.
implies([],_) :- !.
implies([MIN|T],DNF) :-
    imply(MIN,DNF),
    implies(T,DNF).

%%% Definition of implication
%%% DNF |= MIN
imply([],_) :- !.
imply(MIN,DNF) :-
    member(M,DNF), subset(M,MIN), !.
imply(MIN,DNF) :-
    find_subsets(DNF,SDNF),
    reduce(SDNF,RSDNF),
    member(M,RSDNF),
    subset(M,MIN).

find_subsets([],[]) :- !.
find_subsets([G|T],[G|T2]):-
    find_subsets(T,T2).
find_subsets([_|T],T2):-
    find_subsets(T,T2).
```

Algorithm 3. BPMN example coding in Prolog: funnel between tasks

```
?- funnel(n4).
dw—>n4—>do
true
```

---

not a part of the BPMN style, thus their model is not completely BPMN compliant and has another goals then ours.

Dijkman and Van Gorp [6] formalized the BPMN model using graph rewrite rules. However, their goal was also different from ours. They focused on execution semantics. Thus, their approach is suitable for simulation, animation and execution of BPMN models.

Wong and Gibbons [33] defined the semantics of BPMN models using Communicating Sequential Processes, in which a process is a pattern of behavior. They provide only verification of such issues as: hierarchical refinement, partial refinement and hierarchical independence.

7 Concluding Remarks

Our work is a part of an approach in the area of of business processes and rules integration [10]. It constitutes an attempt at providing a logical, declarative model for well-defined BPMN diagram [15]. The model is aimed at defining formal semantics of diagram components and the workflow operation. The main focus is on the specification of correct components and correct dataflow. Global termination conditions are specified in a recursive way. Summarizing, in the paper we:

- formulated a formal model for a subset of BPMN,
- defined local and global correctness requirements,
- specified the detailed logic that stems from the diagram.

The original contribution of our work consists in:

- presenting open issues corresponding to the BPMN diagrams, such as consistency of elements,
- specifying a BPMN model in the declarative Prolog language, which helps to check the defined correctness requirements.

Note that the logical analysis can be performed *off-line*, on the base of logical requirements ϕ, ψ and φ of data. However, if such specifications are data-dependent (e.g. in case of loops or more complex non-monotonic data processing) the analysis may be possible only in *on-line* form, separately for any admissible input data.

As future work, a more complex modeling, verification and execution approach is considered. In the case of modeling issue, we plan to implement this approach by extending one of the existing BPMN tools in order to integrate it with the HeKatE Qt Editor (HQEd) for XTT2-based Business Rules [9]. The XTT2 rules [21] (and tables) can be formally analyzed using the so-called verification HalVA framework [10] or using Petri net approach [31]. Although table-level verification can be performed with HalVA [18], the global verification is a more complex issue [11]. Our preliminary works on global verification have been presented in [30].

Acknowledgment. The paper is supported by the *BIMLOQ* Project funded from 2010–2012 resources for science as a research project.

References

1. Allweyer, T.: BPMN 2.0. Introduction to the Standard for Business Process Modeling. BoD, Norderstedt (2010)
2. Ambler, S.W.: Business Rules (2003),
   http://www.agilemodeling.com/artifacts/businessRule.htm
3. Andersson, B., Bergholtz, M., Edirisuriya, A., Ilayperuma, T., Johannesson, P.: A declarative foundation of process models. In: Pastor, Ó., Falcão e Cunha, J. (eds.) CAiSE 2005. LNCS, vol. 3520, pp. 233–247. Springer, Heidelberg (2005)
4. Andročec, D.: Simulating BPMN models with Prolog. In: Proceedings from the Central European Conference on Information and Intelligent Systems, CECIIS 2010, pp. 363–368 (2010)
5. Dijkman, R.M., Dumas, M., Ouyang, C.: Formal semantics and automated analysis of BPMN process models. preprint 7115. Tech. rep., Queensland University of Technology, Brisbane, Australia (2007)
6. Dijkman, R., Van Gorp, P.: Bpmn 2.0 execution semantics formalized as graph rewrite rules. In: Mendling, J., Weidlich, M., Weske, M. (eds.) BPMN 2010. LNBIP, vol. 67, pp. 16–30. Springer, Heidelberg (2010)
7. Giurca, A., Gašević, D., Taveter, K. (eds.): Handbook of Research on Emerging Rule-Based Languages and Technologies: Open Solutions and Approaches. Information Science Reference, Hershey (2009)
8. Gruhn, V., Laue, R.: Checking properties of business process models with logic programming. In: Augusto, J.C., Barjis, J., Ultes-Nitsche, U. (eds.) Proceedings of the 5th International Workshop on Modelling, Simulation, Verification and Validation of Enterprise Information Systems, MSVVEIS 2007, In conjunction with ICEIS 2007, Funchal, Madeira, Portugal, pp. 84–93. Insticc Press (June 2007)
9. Kluza, K., Kaczor, K., Nalepa, G.J.: Enriching business processes with rules using the Oryx BPMN editor. In: Rutkowski, L., Korytkowski, M., Scherer, R., Tadeusiewicz, R., Zadeh, L.A., Zurada, J.M. (eds.) ICAISC 2012, Part II. LNCS, vol. 7268, pp. 573–581. Springer, Heidelberg (2012),
   http://www.springerlink.com/content/u654r0m56882np77/
10. Kluza, K., Maślanka, T., Nalepa, G.J., Ligęza, A.: Proposal of representing BPMN diagrams with XTT2-based business rules. In: Brazier, F.M.T., Nieuwenhuis, K., Pavlin, G., Warnier, M., Badica, C. (eds.) Intelligent Distributed Computing V. SCI, vol. 382, pp. 243–248. Springer, Heidelberg (2011),
    http://www.springerlink.com/content/d44n334p05772263/
11. Kluza, K., Nalepa, G.J., Szpyrka, M., Ligęza, A.: Proposal of a hierarchical approach to formal verification of BPMN models using Alvis and XTT2 methods. In: Canadas, J., Nalepa, G.J., Baumeister, J. (eds.) 7th Workshop on Knowledge Engineering and Software Engineering (KESE2011) at the Conference of the Spanish Association for Artificial Intelligence (CAEPIA 2011), La Laguna, Tenerife, Spain, pp. 15–23 (November 10, 2011),
    http://ceur-ws.org/Vol-805/
12. Lam, V.S.W.: Equivalences of BPMN processes. Service Oriented Computing and Applications 3(3), 189–204 (2009)

13. Lam, V.S.W.: Formal analysis of BPMN models: a NuSMV-based approach. International Journal of Software Engineering and Knowledge Engineering 20(7), 987–1023 (2010)
14. Ligęza, A.: Logical Foundations for Rule-Based Systems. SCI, vol. 11. Springer, Heidelberg (2006)
15. Ligęza, A.: BPMN – a logical model and property analysis. Decision Making in Manufacturing and Services 5(1-2), 57–67 (2011)
16. Ligęza, A., Kluza, K., Potempa, T.: Ai approach to formal analysis of bpmn models. towards a logical model for bpmn diagrams. In: Ganzha, M., Maciaszek, L.A., Paprzycki, M. (eds.) Proceedings of the Federated Conference on Computer Science and Information Systems, FedCSIS 2012, Wroclaw, Poland, September 9-12, pp. 931–934 (2012), http://ieeexplore.ieee.org/xpls/abs_all.jsp?arnumber=6354394
17. Ligęza, A., Nalepa, G.J.: A study of methodological issues in design and development of rule-based systems: proposal of a new approach. Wiley Interdisciplinary Reviews: Data Mining and Knowledge Discovery 1(2), 117–137 (2011), doi:10.1002/widm.11
18. Nalepa, G.J., Bobek, S., Ligęza, A., Kaczor, K.: HalVA – rule analysis framework for XTT2 rules. In: Bassiliades, N., Governatori, G., Paschke, A. (eds.) RuleML 2011 - Europe. LNCS, vol. 6826, pp. 337–344. Springer, Heidelberg (2011), http://www.springerlink.com/content/c276374nh9682jm6/
19. Nalepa, G.J.: Semantic Knowledge Engineering. A Rule-Based Approach. Wydawnictwa AGH, Kraków (2011)
20. Nalepa, G.J., Kluza, K., Kaczor, K.: Proposal of an inference engine architecture for business rules and processes. In: Rutkowski, L., Korytkowski, M., Scherer, R., Tadeusiewicz, R., Zadeh, L.A., Zurada, J.M. (eds.) ICAISC 2013, Part II. LNCS (LNAI), vol. 7895, pp. 453–464. Springer, Heidelberg (2013), http://www.springer.com/computer/ai/book/978-3-642-38609-1
21. Nalepa, G.J., Ligęza, A.: HeKatE methodology, hybrid engineering of intelligent systems. International Journal of Applied Mathematics and Computer Science 20(1), 35–53 (2010)
22. Negnevitsky, M.: Artificial Intelligence. A Guide to Intelligent Systems. Addison-Wesley, Harlow (2002) ISBN 0-201-71159-1
23. OMG: Production Rule Representation RFP. Tech. rep., Object Management Group (2003)
24. OMG: Semantics of Business Vocabulary and Business Rules (SBVR). Tech. Rep. dtc/06-03-02, Object Management Group (2006)
25. OMG: Business Process Model and Notation (BPMN): Version 2.0 specification. Tech. Rep. formal/2011-01-03, Object Management Group (2011)
26. Ouyang, C., Wil, M.P., van der Aalst, M.D., ter Hofstede, A.H.: Translating BPMN to BPEL. Tech. rep., Faculty of Information Technology, Queensland University of Technology, GPO Box 2434, Brisbane QLD 4001, Australia Department of Technology Management, Eindhoven University of Technolog y, GPO Box 513, NL-5600 MB, The Netherlands (2006)
27. Ouyang, C., Dumas, M., ter Hofstede, A.H., van der Aalst, W.M.: From bpmn process models to bpel web services. In: IEEE International Conference on Web Services, ICWS 2006 (2006)
28. Ross, R.G.: The RuleSpeak Business Rule Notation. Business Rules Journal 7(4) (2006), http://www.BRCommunity.com/a2006/b282.html
29. Silver, B.: BPMN Method and Style. Cody-Cassidy Press (2009)

30. Szpyrka, M., Nalepa, G.J., Ligęza, A., Kluza, K.: Proposal of formal verification of selected BPMN models with Alvis modeling language. In: Brazier, F.M.T., Nieuwenhuis, K., Pavlin, G., Warnier, M., Badica, C. (eds.) Intelligent Distributed Computing V. SCI, vol. 382, pp. 249–255. Springer, Heidelberg (2011), http://www.springerlink.com/content/m181144037q67271/
31. Szpyrka, M., Szmuc, T.: Decision tables in petri net models. In: Kryszkiewicz, M., Peters, J.F., Rybiński, H., Skowron, A. (eds.) RSEISP 2007. LNCS (LNAI), vol. 4585, pp. 648–657. Springer, Heidelberg (2007)
32. White, S.A., Miers, D.: BPMN Modeling and Reference Guide: Understanding and Using BPMN. Future Strategies Inc., Lighthouse Point (2008)
33. Wong, P.Y.H., Gibbons, J.: A process semantics for bpmn. In: Liu, S., Araki, K. (eds.) ICFEM 2008. LNCS, vol. 5256, pp. 355–374. Springer, Heidelberg (2008)

Square Complexity Metrics for Business Process Models*

Krzysztof Kluza, Grzegorz J. Nalepa, and Janusz Lisiecki

Abstract. Complexity metrics for Business Process (BP) are used for the better understanding, and controlling quality of the models, thus improving their quality. In the paper we give an overview of the existing metrics for describing various aspects of BP models. We argue, that the design process of BP models can be improved by the availability of metrics that are transparent and easy to be interpreted by the designers. Therefore, we propose simple yet practical square metrics for describing complexity of a BP model based on the Durfee and Perfect square concept. These metrics are easy to interpret and provide basic information about the structural complexity of the model. The proposed metrics are to be used with models built with Business Process Model and Notation (BPMN), which is currently the most widespread language used for BP modeling. Moreover, we present a set of BPMN models analyzed with our metrics. Finally, we introduce a tool implementing the discussed metrics. We compare the results to other important metrics, emphasizing the qualities of our approach.

1 Introduction and Motivation

Business Process (BP) models constitute a graphical representation of processes in an organization. The Business Process Model and Notation (BPMN) [36] visual language contributed significantly to Software Engineering (SE) in terms of collaboration between developers, software architects and business analysts. Although there are many new tools and methodologies in this area, they do not make BPMN models more comprehensible. Thus, a set of best practices for modelers is needed, and some *methods measuring features of model complexity* are desired.

Krzysztof Kluza · Grzegorz J. Nalepa · Janusz Lisiecki
AGH University of Science and Technology,
al. A. Mickiewicza 30, 30-059 Krakow, Poland
e-mail: {kluza,gjn}@agh.edu.pl

* The paper is supported by the AGH UST 11.11.120.859.

M. Mach-Król and T. Pełech-Pilichowski (eds.), *Advances in Business ICT*,
Advances in Intelligent Systems and Computing 257,
DOI: 10.1007/978-3-319-03677-9_6, © Springer International Publishing Switzerland 2014

To be a base for communication, models should be easy to understand and to maintain. Although visualization can help in evaluation task [3], it would be useful to have *measures that can provide information about understandability and maintainability* of a BP model [23]. This can help to evaluate the difficulty of producing a BP before its implementation. BP metrics can help to control, estimate and improve processes and therefore organizations during design [38]. As it was validated by Mendling in [27], proper complexity measures *have the potential to serve as modeling error probability determinants* [26].

Our motivation is to deliver new complexity metrics for business processes that are easy to grasp for different groups of users. We argue, that the design process of business process models can be improved by the availability of metrics that are transparent and easy to be interpreted.

This paper is an extended version of the paper [19] presented at the ABICT 2012 workshop. It gives an overview of selected existing metrics for describing various aspects of BP models. Primarily we focus on measurements which are easy to compute, i.e. they are intuitive and not difficult to interpret by users. As the original contribution, we propose a new, simple, yet practical metric for BPMN models, which adopts concepts of some commonly known bibliometric indicators for the BP model purposes. We extend the discussion of related works and evaluation of results. Moreover, we present a set of business process models analyzed with our metrics. Finally, we introduce a tool implementing the discussed metrics. We compare the results to other important metrics, emphasizing the qualities of our approach.

The rest of this paper is organized as follows: Section 2 presents BPMN model and its elements. Section 3 describes the existing approaches to measuring various aspects of BP models. In Section 4 the new square metrics for BP are proposed. The paper is summarized in Section 6.

2   Main Elements of the BPMN Model

A Business Process [47] can be defined as a collection of related tasks that produce a specific service or product (serve a particular goal) for a particular customer. BPMN [36] constitutes the most widespread language for modeling BP. It uses a set of predefined graphical elements to depict a process and how it is performed. The current BPMN 2.0 specification defines three models to cover various aspects of processes:

1. *Process Model* – describes the ways in which operations are carried out to accomplish the intended objectives of an organization. The process can be modeled on different abstraction levels: *public* (collaborative Business 2 Business Processes) or *private* (internal Business Processes).
2. *Choreography Model* – defines expected behavior between two or more interacting business participants in the process.

3. *Collaboration Model* – can include Processes and/or Choreographies, and provides a Conversation view (which specifies the logical relation of message exchanges).

Although BPMN 2.0 defines three models to cover various aspects of processes, in most cases, using only the Process Model is sufficient. Four basic categories of elements used to model such processes are: flow objects (*activities, gateways,* and *events*), connecting objects (*sequence flows, message flows,* and *associations*), swimlanes, and artifacts, see Fig. 1. Activities are the main BPMN elements. They denote tasks that have to be performed and are represented by rectangles with rounded corners. The sequence flow between activities, the flow of control, is depicted by arcs. The directions of arcs depict the order in which the activities have to be performed. Events, represented by circles, denote something that happens during the lifetime of the process. The icon within the circle denotes the event type. e.g. envelope for *message event,* clock for *time event.* Gateways, represented by diamond shapes, determine forking and merging of the sequence flow between tasks in a process, depending on some conditions. The icon within the shape denotes the logical function associated with the gateway.

In our study, we focus on the mentioned above main elements of the BPMN process model. For our approach what matters most are the key structural elements that shape the complexity of the analyzed BP models.

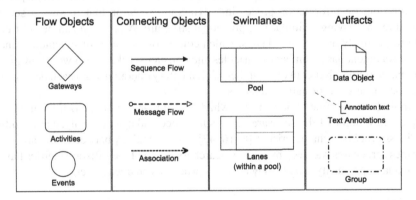

Fig. 1 BPMN core objects

3 State of the Art in the Metrics for Business Processes

One of the most influential papers in the business process modeling field by Mendling et al. [29] concerns guidelines for modelers, which should be taken into account when modeling business processes. They formulated seven guidelines and prioritized them with the help of industry experts [29]:

1. Model as structured as possible.
2. Decompose a model with more than 50 elements.
3. Use as few elements in the model as possible.
4. Use verb-object activity labels.
5. Minimize the routing paths per element.
6. Use one start and one end event.
7. Avoid OR routing elements.

Such principles help with modeling more comprehensive and less error-prone designs. According to Mendling et al., following conclusions can be drawn:

- Smaller BP models are more manageable, with less structural errors.
- There is a connection between nodes degree and the model understandability.
- Model structuredness affects both intelligibility and the number of defects.

However, the existing tools do not require to comply with such requirements, so a user has to adhere to these guidelines itself.

In the past decades, a lot of research has been conducted on the measurement of software programs [24]. Metrics have been used for many purposes, such as predicting errors [44], supporting refactoring or estimating costs of software [14] or measuring software functional size [30]. Khlif et al. noticed the similarity between object-oriented software and business process model [16]. Other authors [37, 43] argue that workflow processes are quite similar to software programs in such respects as: focusing on information processing, dynamic execution that follows a static structure and having a similar compositional structure. A program can be split up into respectively modules or classes, which consist of a number of statements, and every statement has a number of variables and constants. Likewise, a workflow process has activities, built out of a number of elementary operations and each operation uses one or more information elements.

Table 1 shows some basic concepts which are similar in both the BPMN and OO approaches. Some of the software metrics have been adapted to analyze and study business processes characteristics [8, 16, 42]. As the analogy between software and business processes was seen by many researchers, some of the software metrics have been adapted to analyze and study business processes characteristics.

Table 1 Similarities between BPMN and OO core concepts [16, 42]

Object oriented software	BPMN models
Classes/Packages	Sub Processes, Processes
Methods	Tasks
Method invocations	Control flow or message flow incoming to a task
Variables/Constants	Data objects
Comment lines	Annotations

According to Conte et al. [9], the quality of software design is related to five design principles: coupling, cohesion, complexity, modularity and size [42]. These principles can be defined as follows:

- *Coupling* is measured by the number of interconnections among modules.
- *Cohesion* is a measure of the relationships of the elements within a module.
- *Complexity* measures the number and size of control constructs.
- *Modularity* measures if the system's components may be separated and recombined by logical partitioning.
- *Size* measures the overall dimension of software.

Most researchers focus on metrics that measure the complexity of the process. IEEE Standard Computer Dictionary defines complexity as *the degree to which a system or component has a design or implementation that is difficult to understand and verify* [13].

As there is no single metric that can be used to measure the complexity of a process, Cardoso et al. distinguished four perspectives to process complexity [6]:

- activity complexity – affected by the number of activities a process has,
- control-flow complexity – affected by such elements and constructs as splits, joins, loops, start and end,
- data-flow complexity – concerns data structures, and the number of formal parameters of activities, and the mappings between data of activities [5],
- resource complexity – concerns different types of resources that have to be accessed during process execution.

In [8], Cardoso et al. introduced several simple complexity metrics adapted from software engineering. Based on the simple metric, which counts the number of Lines of Code (**LOC**) of a program, they proposed three metrics:

- **NOA** – Number of activities in a process.
- **NOAC** – Number of activities and control-flow elements.
- **NOAJS** – Number of activities, joins, and splits.

To evaluate the difficulty of producing business process, Cardoso introduced Control-Flow Complexity (**CFC**) metric, which borrows some ideas from McCabe's cyclomatic complexity [7]. This metric uses the number of states induced by control-flow elements in a process. Moreover, for estimating process length, volume, and difficulty, Cardoso et al. introduced the notion of Halstead-based Process Complexity (**HPC**) measures [8]. This composite measure of complexity comprises a set of primitive measures:

- $n1$ – the number of unique activities, splits and joins, and other control-flow elements,
- $n2$ – the number of unique data variables that are manipulated by the process and its activities,
- $N1$ and $N2$ can be easily derived directly from $n1$ and $n2$ as total numbers of the control flow and data elements.

Thus, the HPC measures can be calculated as follows:

- **process length:** $N = n1 * log_2(n1) + n2 * log_2(n2)$
- **process volume:** $V = (N1 + N2) * log_2(n1 + n2)$
- **process difficulty:** $D = (n1/2) * (N2/n2)$

Although the HPC measures for processes can have several advantages (do not require in-depth analysis of process structures and can predict rate of errors) [8], it is not defined how the primitive measures should be adapted for the purpose of the BPMN notation.

Another metric adapted by Cardoso et al. is the Information Flow Metric. They proposed a metric Interface Complexity (**IC**) of an activity, defined as: $IC = Length * (NoI * NoO)^2$. In the formula, the length of the activity can be calculated using traditional SE metrics such as LOC (Cardoso et al. suggested 1 if the activity source code is unknown) and the number of inputs (**NoI**) and outputs (**NoO**) of the activity follow directly from the model.

Latva-Koivisto [22] proposed the Coefficient of Network Complexity (**CNC**) metric for business processes. CNC is a widely used metric in network analysis and was proposed to measure the degree of complexity of a critical pass network. Cardoso et al. [8] proposed a more precise version of the formula: $CNC = number of arcs / (no. of activities, joins, and splits)$.

Abreu et al. specified the presented metrics (CNC, KNH, CFC, HPC) with the Object Constraint Language (OCL) upon a lightweight BPMN metamodel [1]. Rolon et al. defined in [2] a large set of simple metrics for the evaluation of business processes at a conceptual level. Their metrics measure the structural complexity of software process models and are grouped into two main categories: Base and Derived Measures. The base measures are defined as a number of each kind of BPMN elements that a business process model is composed of, e.g. **NT** (Number of Tasks) – the total number of tasks in a process model. Based on such measures they proposed several derived measures which uses some simple measurement function that describes the proportions among different elements of the model, e.g. **TNA/NSFA** (Total Number of Activities/Number of Sequence Flows between Activities) – describes connectivity level between activities in a process.

Selected simple metrics defined by Sánchez-González et al. [39] are presented in Table 2. Each of these metrics is easy to understand and describes a single aspect of a process model.

In [23], Gruhn and Laue took into consideration other properties of processes to define some new metrics. They proposed analyzed nesting properties – Maximum Nesting Depth (**MaxND**) and Mean Nesting Depth (**MeanND**): The nesting depth of an action is the number of decisions in the control flow that are necessary to perform this action. They noticed that a greater nesting depth implies greater complexity and both nesting depth metrics have a strong influence on other structure-related complexity metrics.

They also formulated Cognitive Complexity (**CC**) metric which uses the cognitive weight for business process models. Their metric is based on the research by Shao and Wang upon the metric which measures the effort required for comprehending certain software. The Cognitive Functional Size (**CFS**) metric for software measurement uses predefined (based on empirical studies) cognitive weights for basic control structures. The cognitive weight of a control structure is a measure for the difficulty to understand this control structure. Hence, such a metric measures indirectly difficulty to understand a model (thus complexity).

Table 2 Simple metrics defined in [39]

Metric	Description
Number of nodes	Number of activities and routing elements in a model.
Diameter	Length of the longest path from a start node to an end node.
Density	Ratio of the total number of arcs to the maximum number of arcs.
The Coefficient of Connectivity	Ratio of the total number of arcs in a process model to its total number of nodes.
The Average Gateway Degree	Average of the number of both incoming and outgoing arcs of the gateway nodes in the process model.
The Maximum Gateway Degree	Maximum sum of incoming and outgoing arcs of these gateway nodes.
Separability	Ratio of the number of cut-vertices on the one hand to the total number of nodes in the process model on the other.
Sequentiality	Degree to which the model is constructed out of pure sequences of tasks.
Depth	Maximum nesting of structured blocks in a process model.
Gateway Mismatch	Sum of gateway pairs that do not match with each other, e.g. when an AND-split is followed by an OR-join.
Gateway Heterogeneity	Number of different types of gateways used in a model.
Cyclicity	Number of nodes in a cycle to the sum of all nodes.
Concurrency	Maximum number of paths in a process model that may be concurrently activate due to AND-splits and OR-splits.

In [37] Reijers and Vanderfeesten defined cohesion and coupling metrics for the design of activities in a workflow design, based on an information processing perspective on workflow processes. Their metrics concerned the relations between activities and information elements, as well as connections between activities in a process. Then in [43], they evaluated their workflow process designs using these metrics. Unfortunately, they do not use the BPMN notation for their solution and the presented metrics are very complex.

Lassen and van der Aalst proposed three metrics for workflow nets [21]:

- Extended Cardoso Metric (**ECaM**) – a Petri net version of metric that generalizes and improves the original CFC metric proposed by Cardoso. It focuses on the syntax of the model and ignore the complexity of the behavior.
- Extended Cyclomatic Metric (**ECyM**) – directly adapted from McCabe Cyclomatic. It focuses on the resulting behavior and ignore the complexity of the model itself.
- Structuredness Metric (**SM**) – which stems from the observation that workflows are often structured in terms of design patterns.

ECaM and ECyM metrics are simple modifications of the CFC metric. SM, in turn, constitutes a new metric, which recognizes different kinds of structures and scores them. In contrast to ECaM and ECyM that focus on a single aspect, behavior or syntax, and do not consider the interaction between different elements. Although

SM can be easily computed by a machine, it is not very intuitive for a designer. Moreover, it concerns Petri net, thus can not be directly used for BPMN. Thus, let us recall some simple metrics which are both intuitive and feasible to measure by human.

In [26] Mendling formalized process metrics with reference to model *partitianability* and *connector interplay*:

- Separability (Π) – it is based on the cut-vertex concept. When the cut-vertex node is removed from the model, the process is split to multiple valid elements. Π metric is defined as the number of cut-vertices divided by the number of all process nodes. Higher separability indicates a less complex model.
- Sequentiality (Ξ) – this metric is defined as a number of connections between non-gateway nodes divided by the total number of connections. If the whole model is sequential, the Ξ ratio is 1. Sequentiality is based on the assumption that models with more gateways are more complex and have more errors.
- Structuredness (Φ) – it is related to simple reduction rules, defined by Mendling in [28]. Metric is calculated as a one minus the number of nodes from the model after reduction, divided by the number of nodes from the original graph. If BP model is well structured, it is considered to be less complex.
- Depth (Λ) – to calculate the depth of a node, one has to take a minimum from the in-depth λ_{in} and the out-depth λ_{out}. Node in-depth is a maximum from difference between split and join gateways, which have to be visited to activate such node. Λ metric can be calculated as a maximum depth of all process nodes. Mendling states that models with higher depth are considered to be less understandable.
- Connector mismatch (**MM**) – with relation to BPMN, proposed metric shows the ratio of mismatched split and join gateways. If mismatch is high, there is a higher probability that the process contains errors, i.e. deadlocks or lack of synchronization.
- Connector heterogeneity (**CH**) – it is a metric which shows how many different gateway types was used in the model. Higher heterogeneity indicates higher probability of connector mismatch. If only one type of gateway was used, the **CH** is 1.
- Cyclicity (**CYC**) – this metric shows how many nodes are in cycle, in relation to all number of nodes in the model. Higher cyclicity signifies less comprehensible model.
- Token Split (**TS**) – measurement is used to evaluate the degree of model parallelism. It is calculated as an output degree of all AND and OR join gateways minus one. Higher token split shows that there is a greater probability of error in process model.

BP model complexity cannot be directly determined by only one type of metric. Mendling shows limitations of various measurements in [26] e.g.

- size, diameter – bigger models can be more understandable than smaller ones if they are more sequential.

- density – there is a high probability that larger (and more complex) BPs are less dense than smaller ones. This metric is insufficient if compared models differ in the number of nodes.
- CNC – one can find a models with the same value of CNC, which vary in comprehensibility due to different types of nodes.
- gateway degree, structuredness, depth, cyclicity – two models which differ in size can have the same value of gateway degree, structuredness, depth and cyclicity.
- separability – this metric will be low if there are many sequential, but parallel nodes in the model.
- sequentiality – models with the same value of sequentiality can vary in complexity due to different kind of gateways.
- token split – understandability of two models with the same token split can differ due to size, structuredness etc.

In several survey papers [41, 31, 4, 10, 20] overviews of the current state of the art in various areas of business process metrics are presented.

Thammarak states in [41], that there is a need for a new metric in relation to process model reusability. Such measurement can be further used by BP designers and improve understandability and manageability. According to author, the metric should mainly be based on model complexity. Thus a few commonly known complexity metrics are briefly described in the survey (LOC, MCC, CFC, HCM). Future aim for the author is to develop appropriate reuse metric, based on the current state of the knowledge.

A Survey by Muketha et al. [31] focuses on complexity metrics. Authors recall, that the three steps need to be done for a new metric to be properly validated:

1. Metric definition,
2. Theoretical validation,
3. Empirical validation.

As for new measurement definition, compliance with the theory of measurement should be met. The first step is to chose of entity and its measurable attributes. Based on that, a new metric can be proposed. For theoretical validation, Muketha et al. propose methods like checking the metric with Weyuker's properties. Empirical validation is based on surveys, case studies and experiments. After a brief description authors summarize well known BP complexity metrics in respects to tools support and theirs theoretical and empirical validation results.

Process model similarity is an intensive field of investigation [11]. Becker and Laue [4] provide a exhaustive survey on BP models similarity metrics. Such measures are based on various properties of BPs:

1. Correspondence of nodes and edges,
2. Graph edit distances,
3. Causal dependencies between activities, and
4. Sets of traces or logs comparison.

Each metric in the paper is precisely described and calculated on example models. After each evaluation, there is a discussion about metric usability for different cases. There are also other papers on process similarity aspects [10, 20]. Proposed and analyzed metrics are related on both model structural and behavioral properties.

It can be concluded that this field of research is still growing and there is a lot of potential for further development of business process metrics. However, there are some signs of immaturity, e.g. there is no single definition how to measure complexity – by coupling metric (Mendling) or size (Gruhn, Laue, and Cardoso et al) [8, 23].

In a systematic review of measurement in business processes [38], Gonzales et al. distinguished several measurable concepts for business process models, such as: complexity, understandability, quality, entropy, density, cohesion and coupling. However, most research in the area of business process measurement focus on complexity, because it is connected with understandability of the process.

This gives us motivation to propose new complexity metrics for business processes that are easy to interpret for different groups of users, and can be easily implemented and used during the design of business process models.

4   Complexity Metrics Based of Durfee and Perfect Square

Although simple metrics, such as NOA or Diameter, can be easily calculated, they do not take into account the variety of structures used by the model. In turn, several metrics that take into account model structure, e.g. CNC or TNA/NSFA, are either complex to calculate or not intuitive for a designer. Moreover, some of presented metrics do not concern the BPMN notation, thus they have to be adapted for this purpose (e.g. HPC).

Based on the simple measures, we propose using a measure which takes into account both types of process elements and their number. The idea originates in such concepts as h-index [15] or Durfee Square[1]. Based on the distribution of the types of process elements: *Durfee Square Metric (DSM) equals d if there are d types of elements which occur at least d times in the model (each), and the other types occur no more than d times (each).*

To give a more accurate representation of the distribution shape, we propose using Perfect Square Metric (PSM) based on the g-index [12] as well. Thus, PSM can be defined as follows: given a set of element types ranked in decreasing order of the number of their instances, the PSM is the (unique) largest number such that the top p types occur (together) at least p^2 times.

Both DSM and PSM are intended to measure simultaneously the variety and the number of process elements. There are several advantages of the proposed metrics. They are very intuitive for a designer (a natural number that is easy to interpret) and not very complex to calculate. Moreover, they can be easily used to measure any process model, particularly for BPMN models.

---

[1] See http://mathworld.wolfram.com/DurfeeSquare.html

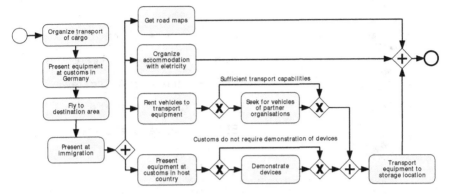

Fig. 2 An exemplary process model for our metrics evaluation [45]

Table 3 Selected metrics calculated for the exemplary process model

Metric	Result for the example
NOA(C/JS)	NOA=11, NOAC=NOAJS=20
CFC	5
HPC	N=V=58.6, D=7.5
CNC	1.2
TNA/NSFA	0.58
Diameter	13
DSM and PSM	DSM=3, PSM=4

Let us consider an exemplary process and calculate the presented metrics for its model. Figure 2 presents a BPMN process model from the domain of earthquake response (presented originally in [45]). Table 3 presents the selected metrics calculated for this exemplary model. Figure 3 shows how the DSM metric was calculated.

An prototype tool that calculates these metrics was implemented. After the preliminary discussion with the BP designers, it can be observed, that proposed measures can be helpful in describing both the complexity of the models, as well as and their understandability and maintainability. The main use scenario for the measures is to help the designer to adapt the designed model and control its complexity by using the calculated measures during the design.

We used this tool on a number of cases to evaluate our metrics as we discuss next.

5 Evaluation

To evaluate the presented metrics, a number of BP cases were evaluated[2]. The cases represented different levels of complexity w.r.t. BPMN artifacts or structures used

---

[2] See http://geist.agh.edu.pl/pub:projects:bimloq:start#cases

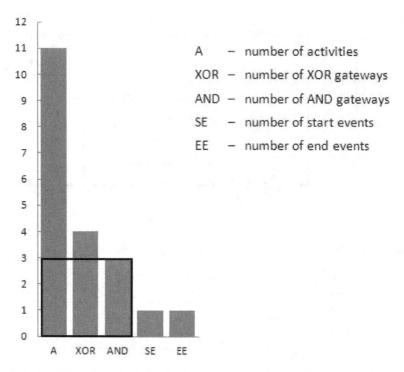

Fig. 3 A chart illustrating calculation for the proposed metric

(selected examples of the cases can be seen in Table 4). The results for selected subset of representative metrics are presented in Table 5.

Based on the values from the Table 5, correlations between the metrics were measured. Result of these calculations are shown in Tables 6 and 7, which present Kendall tau rank and Spearman's correlation coefficient. Each rank is computed by Multivariate Correlation Matrix (v1.0.5) [46].

Metric rank correlation computed for example cases are presented in the Figures 4 and 5. The data series histogram is shown in the matrix diagonal. The smooth curve and scatterplots for every combination of pairs of data series are presented in upper half of the matrix. There is also a number in lower half that represents the p-value of the (Kendall tau / Spearman) correlation. The p-value is used to test and reject the null hypothesis. The correlation is not considered to be a result of random data if the p-value is less than given *significance level*, often 0.05. For every scatterplot (in the upper half) there is a corresponding p-value in the lower half. The name of each metric is displayed on the diagonal.

Table 4 The exemplary process model cases used for metrics evaluation

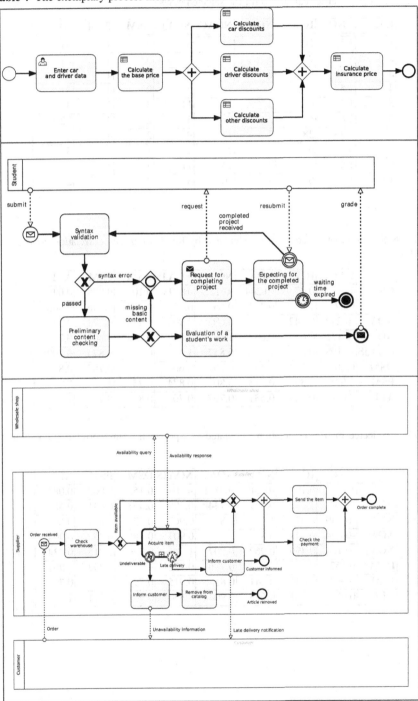

Table 5 Selected metrics calculated for the selected cases

Case	CNC	IC	NOA	NOAC	NOAJS	DSM	PSM	ALL
1	1.5	20	5	13	8	3	4	31
2	1.38	20	5	11	8	3	4	28
3	1.56	7	6	10	8	2	4	24
4	1.57	16	4	10	7	2	4	21
5	1.45	63	7	17	11	3	5	43
6	2.67	8	2	7	3	2	3	22
7	1.33	28	7	13	9	2	4	26
8	0.5	0	2	2	2	1	1	3
9	1.45	7	7	13	11	4	5	38
10	1.38	6	6	10	8	2	4	21

Table 6 Selected metrics correlation calculated by Kendall tau rank correlation coefficient [46]

	CNC	IC	NOA	NOAC	NOAJS	DSM	PSM	ALL
CNC	1	0.07	-0.244	-0.122	-0.198	0.108	0.028	0.046
IC	0.07	1	0.342	0.659	0.47	0.404	0.396	0.506
NOA	-0.244	0.342	1	0.692	0.909	0.453	0.773	0.531
NOAC	-0.122	0.659	0.692	1	0.857	0.708	0.803	0.797
NOAJS	-0.198	0.47	0.909	0.857	1	0.66	0.874	0.709
DSM	0.108	0.404	0.453	0.708	0.66	1	0.755	0.8
PSM	0.028	0.396	0.773	0.803	0.874	0.755	1	0.7
ALL	0.046	0.506	0.531	0.797	0.709	0.8	0.7	1

Table 7 Selected metrics correlation calculated by Spearman's rank correlation coefficient [46]

	CNC	IC	NOA	NOAC	NOAJS	DSM	PSM	ALL
CNC	1	0.104	-0.249	-0.094	-0.164	0.152	0.062	0.083
IC	0.104	1	0.377	0.744	0.499	0.468	0.441	0.661
NOA	-0.249	0.377	1	0.8	0.962	0.523	0.841	0.668
NOAC	-0.094	0.744	0.8	1	0.907	0.786	0.85	0.904
NOAJS	-0.164	0.499	0.962	0.907	1	0.726	0.915	0.834
DSM	0.152	0.468	0.523	0.786	0.726	1	0.793	0.901
PSM	0.062	0.441	0.841	0.85	0.915	0.793	1	0.791
ALL	0.083	0.661	0.668	0.904	0.834	0.901	0.791	1

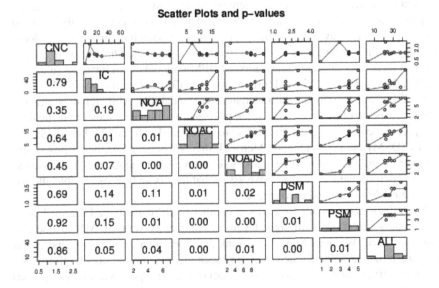

Fig. 4 Scatter plots and P–values of Kendall tau correlation coefficient [46]

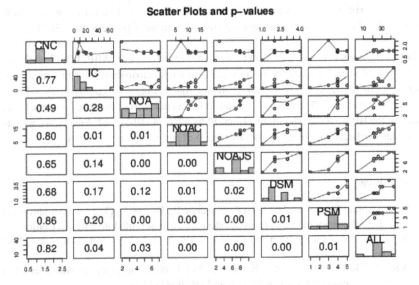

Fig. 5 Scatter plots and P–values of Spearman correlation coefficient [46]

6 Conclusions and Future Works

This paper focuses on the problem of analyzing the business process characteristics and measuring its quality usign metrics. Currently, a lot of research is carried out in the area of business processes measurement. Most of the initiatives in this area are adapted from the software engineering field and have not been empirically validated yet. We provided an overview of the current state of research in this area, in order to identify existing research gaps and future research directions. This fits to our current research directions covering integration of BPMN models with business rules [32, 18, 25], their execution [17] and formal modeling [35] and verification [40].

Complex metrics struggle with human understandability. Although they measure different aspects of processes and often provide valuable data, they are difficult to interpret. To cope with these problems we introduced two metrics Durfee Square Metric (DSM) and Perfect Square Metric (PSM). We believe that the proposed simple metrics, are very intuitive and easy to explain to users, since they expressed as easy to interpret natural numbers, We assert that they 0can be a reasonable alternative to already known and used metrics.

To evaluate the presented metrics, a number of BP cases have been evaluated. These cases represented different levels of complexity w.r.t. BPMN artifacts or structures used. We used Kendall tau rank and Spearman's correlation coefficient to characterize the selected metrics. Using such metrics one can obtain information about the structural complexity of the model of business processes. It is important to mention that the proposed metrics are not difficult to calculate. This would allow for quick comparison of two different models, and evaluation of their quality at a conceptual level. The research presented in this paper is a proposal for further studies. The experimental tool is also being integrated as a part of an online wiki system allowing for the collaborative authoring end evaluation of process models [33, 34]

Acknowledgements. The authors wish to thank the organizers of the ABICT 2012 workshop on the FedCSIS 2012 conference for providing an ample environment for presenting and discussing our research. We also thank professor Antoni Ligęza for giving many valuable comments on our work. The methods presented in this paper were evaluated using selected results of the BIMLOQ research project funded by NCN in 2010–2012. Finally, we wish to thank Mateusz Baran for providing the description of business process models.

References

1. Brito e Abreu, F., de Braganca V da Porciuncula, R., Freitas, J., Costa, J.: Definition and validation of metrics for itsm process models. In: 2010 Seventh International Conference on the Quality of Information and Communications Technology (QUATIC), pp. 79–88 (2010)
2. Aguilar, E.R., Ruiz, F., García, F., Piattini, M.: Applying software metrics to evaluate business process models. CLEI Electronic Journal 9(1) (2006)
3. Baumeister, J., Freiberg, M.: Knowledge visualization for evaluation tasks. Knowledge and Information Systems 29(2), 349–378 (2011)
4. Becker, M., Laue, R.: A comparative survey of business process similarity measures. Computers in Industry 63(2), 148–167 (2012)

5. Cardoso, J.: About the data-flow complexity of web processes. In: Proceedings from the 6th International Workshop on Business Process Modeling, Development, and Support: Business Processes and Support Systems: Design for Flexibility, The 17th Conference on Advanced Information Systems Engineering, CAiSE 2005, June 13-17, Porto, Portugal, pp. 67–74 (2005)
6. Cardoso, J.: Control-flow complexity measurement of processes and weyuker's properties. In: 6th International Enformatika Conference. Transactions on Enformatika, Systems Sciences and Engineering, Budapest, Hungary, October 26-28, vol. 8, pp. 213–218 (2005)
7. Cardoso, J.: How to measure the control-flow complexity of web processes and workflows. In: Fischer, L. (ed.) Workflow Handbook 2005, pp. 199–212. Future Strategies Inc., Lighthouse Point (2005)
8. Cardoso, J., Mendling, J., Neumann, G., Reijers, H.A.: A discourse on complexity of process models. In: Eder, J., Dustdar, S. (eds.) BPM Workshops 2006. LNCS, vol. 4103, pp. 117–128. Springer, Heidelberg (2006)
9. Conte, S.D., Dunsmore, H.E., Shen, V.Y.: Software engineering metrics and models. Benjamin-Cummings Publishing Co. Inc., Redwood City (1986)
10. Dijkman, R., Dumas, M., van Dongen, B., Käärik, R., Mendling, J.: Similarity of business process models: Metrics and evaluation. Information Systems 36(2), 498–516 (2011)
11. Dijkman, R.M., Dongen, B.F., Dumas, M., Garcia-Banuelos, L., Kunze, M., Leopold, H., Mendling, J., Uba, R., Weidlich, M., Weske, M., Yan, Z.: A short survey on process model similarity. In: Bubenko, J., Krogstie, J., Pastor, O., Pernici, B., Rolland, C., Solvberg, A. (eds.) Seminal Contributions to Information Systems Engineering, pp. 421–427. Springer, Heidelberg (2013)
12. Egghe, L.: Theory and practise of the g-index. Scientometrics 69(1), 131–152 (2006)
13. Geraci, A.: IEEE Standard Computer Dictionary: Compilation of IEEE Standard Computer Glossaries. IEEE Press (1991)
14. Grady, R.: Successfully applying software metrics. Computer 27(9), 18–25 (1994)
15. Hirsch, J.E.: An index to quantify an individual's scientific research output. PNAS 102(46), 16,569–16,572 (2005)
16. Khlif, W., Zaaboub, N., Ben-Abdallah, H.: Coupling metrics for business process modeling. International Journal of Computers 4(4) (2010)
17. Kluza, K., Kaczor, K., Nalepa, G.J.: Enriching business processes with rules using the Oryx BPMN editor. In: Rutkowski, L., Korytkowski, M., Scherer, R., Tadeusiewicz, R., Zadeh, L.A., Zurada, J.M. (eds.) ICAISC 2012, Part II. LNCS, vol. 7268, pp. 573–581. Springer, Heidelberg (2012),
http://www.springerlink.com/content/u654r0m56882np77/
18. Kluza, K., Maślanka, T., Nalepa, G.J., Ligęza, A.: Proposal of representing BPMN diagrams with XTT2-based business rules. In: Brazier, F.M.T., Nieuwenhuis, K., Pavlin, G., Warnier, M., Badica, C. (eds.) Intelligent Distributed Computing V. SCI, vol. 382, pp. 243–248. Springer, Heidelberg (2011),
http://www.springerlink.com/content/d44n334p05772263/
19. Kluza, K., Nalepa, G.J.: Proposal of square metrics for measuring business process model complexity. In: Ganzha, M., Maciaszek, L.A., Paprzycki, M. (eds.) Proceedings of the Federated Conference on Computer Science and Information Systems, FedCSIS 2012, Wroclaw, Poland, September 9-12, pp. 919–922 (2012),
http://ieeexplore.ieee.org/xpls/abs_all.jsp?
arnumber=6354395

20. Kunze, M., Weidlich, M., Weske, M.: Behavioral similarity – A proper metric. In: Rinderle-Ma, S., Toumani, F., Wolf, K. (eds.) BPM 2011. LNCS, vol. 6896, pp. 166–181. Springer, Heidelberg (2011)
21. Lassen, K.B., van der Aalst, W.M.P.: Complexity metrics for workflow nets. Information and Software Technology 51(3), 610–625 (2009)
22. Latva-Koivisto, A.M.: Finding a complexity for business process models. Tech. rep., Helsinki University of Technology (2001)
23. Laue, R., Gruhn, V.: Complexity metrics for business process models. In: Witold Abramowicz, H.C.M. (ed.) Business Information Systems, 9th International Conference on Business Information Systems, BIS 2006, Klagenfurt, Austria, May 31-June 2, pp. 1–12 (2006)
24. Ligęza, A.: Intelligent data and knowledge analysis and verification; towards a taxonomy of specific problems. In: Vermesan, A., Coenen, F. (eds.) Validation and Verification of Knowledge Based Systems: Theory, Tools and Practice, pp. 313–325. Kluwer Academic Publishers (1999)
25. Ligęza, A., Nalepa, G.J.: A study of methodological issues in design and development of rule-based systems: proposal of a new approach. Wiley Interdisciplinary Reviews: Data Mining and Knowledge Discovery 1(2), 117–137 (2011), doi:10.1002/widm.11
26. Mendling, J.: Metrics for business process models. In: Mendling, J. (ed.) Metrics for Process Models. LNBIP, vol. 6, pp. 103–133. Springer, Heidelberg (2009)
27. Mendling, J.: Validation of metrics as error predictors. In: Metrics for Process Models. LNBIP, vol. 6, pp. 135–150. Springer, Heidelberg (2009)
28. Mendling, J.: Verification of epc soundness. In: Metrics for Process Models. LNBIP, vol. 6, pp. 59–102. Springer, Heidelberg (2009)
29. Mendling, J., Reijers, H.A., van der Aalst, W.M.P.: Seven process modeling guidelines (7pmg). Information & Software Technology 52(2), 127–136 (2010)
30. Monsalve, C., Abran, A., April, A.: Measuring software functional size from business process models. International Journal of Software Engineering and Knowledge Engineering 21(3), 311–338 (2011)
31. Muketha, G., Ghani, A.A.A., Selamat, M.H., Atan, R.: A survey of business process complexity metrics. Information Technology Journal 9(7), 1336–1344 (2010)
32. Nalepa, G.J.: Proposal of business process and rules modeling with the XTT method. In: Negru, V., et al. (eds.) SYNASC Ninth International Symposium Symbolic and Numeric Algorithms for Scientific Computing, September 26-29, pp. 500–506. IEEE Computer Society, IEEE, CPS Conference Publishing Service, Los Alamitos (2007)
33. Nalepa, G.J.: PlWiki – a generic semantic wiki architecture. In: Nguyen, N.T., Kowalczyk, R., Chen, S.-M. (eds.) ICCCI 2009. LNCS (LNAI), vol. 5796, pp. 345–356. Springer, Heidelberg (2009)
34. Nalepa, G.J.: Collective knowledge engineering with semantic wikis. Journal of Universal Computer Science 16(7), 1006–1023 (2010),
    http://www.jucs.org/jucs_16_7/collective_knowledge_
    engineering_with
35. Nalepa, G.J., Ligęza, A., Kaczor, K.: Formalization and modeling of rules using the XTT2 method. International Journal on Artificial Intelligence Tools 20(6), 1107–1125 (2011)
36. OMG: Business Process Model and Notation (BPMN): Version 2.0 specification. Tech. Rep. formal/2011-01-03, Object Management Group (2011)
37. Reijers, H., Vanderfeesten, I.: Cohesion and coupling metrics for workflow process design. In: Desel, J., Pernici, B., Weske, M. (eds.) BPM 2004. LNCS, vol. 3080, pp. 290–305. Springer, Heidelberg (2004)

38. Sánchez-González, L., García, F., González, F.R., Velthuís, M.P.: Measurement in business processes: a systematic review. Business Process Management Journal 16(1), 114–134 (2010)
39. Sánchez-González, L., García, F., Mendling, J., Ruiz, F., Piattini, M.: Prediction of business process model quality based on structural metrics. In: Parsons, J., Saeki, M., Shoval, P., Woo, C., Wand, Y. (eds.) ER 2010. LNCS, vol. 6412, pp. 458–463. Springer, Heidelberg (2010)
40. Szpyrka, M., Nalepa, G.J., Ligęza, A., Kluza, K.: Proposal of formal verification of selected BPMN models with Alvis modeling language. In: Brazier, F.M.T., Nieuwenhuis, K., Pavlin, G., Warnier, M., Badica, C. (eds.) Intelligent Distributed Computing V. SCI, vol. 382, pp. 249–255. Springer, Heidelberg (2011), http://www.springerlink.com/content/m181144037q67271/
41. Thammarak, K.: Survey complexity metrics for reusable business process. In: Proceedings from 1st National Conference on Applied Computer Technology and Information System, ACTIS 2010, pp. 18–22. Bangkok Suvarnabhumi College (2010)
42. Vanderfeesten, I., Cardoso, J., Mendling, J., Reijers, H., van der Aalst, W.: Quality metrics for business process models. In: Fischer, L. (ed.) BPM and Workflow Handbook 2007, pp. 179–190. Future Strategies Inc., Lighthouse Point (2007)
43. Vanderfeesten, I., Reijers, H.A., van der Aalst, W.M.P.: Evaluating workflow process designs using cohesion and coupling metrics. Computers in Industry 59(5), 420–437 (2008)
44. Wang, H., Khoshgoftaar, T.M., Hulse, J.V., Gao, K.: Metric selection for software defect prediction. International Journal of Software Engineering and Knowledge Engineering 21(2), 237–257 (2011)
45. Weidlich, M., Zugal, S., Pinggera, J., Fahland, D., Weber, B., Reijers, H., Mendling, J.: The impact of change task type on maintainability of process models. In: Proceedings from the 1st Workshop on Empirical Research in Process-Oriented Information Systems (ER-POIS 2010), Tunesia, June 7-8, pp. 43–54 (2010)
46. Wessa, P.: Multivariate correlation matrix (v1.0.4) in free statistics software (v1.1.23-r6) (2010), http://www.wessa.net/Patrick.Wessa/rwasp_pairs.wasp/, http://www.wessa.net/Patrick.Wessa/rwasp_pairs.wasp/
47. White, S.A., Miers, D.: BPMN Modeling and Reference Guide: Understanding and Using BPMN. Future Strategies Inc., Lighthouse Point (2008)

Spatial Component in Business Intelligence System for Advanced Threat and Risk Analysis

Mirza Ponjavic and Almir Karabegovic

Abstract. This paper shows an innovative approach for implementation business intelligence systems in advanced threat and risk analysis using spatial component. It demonstrates how to improve intelligence of complete information system by involving spatial extension. Most of business data in data warehouses are often spatial per se, and without using this component, analysis missing very important dimension of the data nature. From other side, frequent problem in enterprise data warehouse is creating relations between tables which come from different sources and without any common attributes; that could be very easily solved by spatial relations. This paradigm of spatialization assumes changing overall system architecture, from data storage, via retrieving to its presentation mechanism. Particular benefit of this approach for threat and risk analysis is effective utilization of location data, advanced spatial analysis techniques and more variety in data visualization. Examples of organizations which need such system are intelligence agencies, emergence services or epidemiology centers.

1 Introduction

Business Intelligence (BI) mainly considers computer-based techniques to support better business decision-making [1]. It uses operations for identifying, extracting, and analyzing business data and offers functions of online analytical processing (OLAP), analytics, data mining, predictive analytics and reporting [2][3]. Business Intelligence data and analysis has more and more importance for business development, but while most of business data has location as a component, few businesses take full advantage of spatial and location analysis.

Location can be described by an address, a geographic region or a tracking route, that can be presented, managed and analyzed interactively in a GIS [4]. Recognized spatial relationships, patterns and trends can answer the sophisticated

Mirza Ponjavic
University of Sarajevo, Faculty of Civil Engineering, Department of Geodesy,
Bosnia and Herzegovina

Almir Karabegovic
University of Sarajevo, Faculty of Electrical Engineering,
Department for Computer Science and Informatics, Bosnia and Herzegovina

M. Mach-Król and T. Pełech-Pilichowski (eds.), *Advances in Business ICT*,
Advances in Intelligent Systems and Computing 257,
DOI: 10.1007/978-3-319-03677-9_7, © Springer International Publishing Switzerland 2014

questions related to hardly visible and invisible laws applied between phenomena described by specific business data sets [5]. Location awareness can be incorporated into Business Intelligence Systems (BIS) and used in field of risk assessment and risk management extracting the maximum value from GIS and BI integration and traditional and spatial data sets.

2   Spatial Component Improves Business Intelligence of Information Systems

Spatial component improves business intelligence of information systems and brings easy-to-understand visualization to business applications. Most of business data in data warehouses are often spatial per se, and without using this component, analysis missing very important dimension of the data nature. From other side, frequent problem in enterprise data warehouse is creating relations between tables which come from different sources and without any common attributes; that could be very easily solved by spatial relations [6]. This paradigm of spatialization assumes changing overall system architecture, from data storage, via retrieving to its presentation mechanism.

Examples of organizations which need such system are risk analysis centers (RAC), various intelligence agencies, emergence services or epidemiology centers. This paper is focused on the implementation of this system for the purpose of improving the center for the analysis of threats and risks, although the experience gained in its implementation may be fully applied in other domains mentioned. Threat and risk analysis is the process used to obtain quantitative or qualitative measures of risk levels and has focus on quantifying the probability of negative consequences from one or more identified or unknown threat causes.

3   Architecture of Business Intelligence System with Spatial Extension for Risk Analysis

An overview of key architectural components of Business Intelligence System (BIS) with spatial extension for RAC project is shown in Fig. 1. Data warehouse (DWH) is intended to store a large amount of data and considers load strategy involving: extracting data from data sources (operational systems), moving it into data warehouse structures, structuring the data for analysis purposes, and moving it into reporting structures (data marts). The architecture includes the process required to handle and manage the following daily operations: data acquisition, data buffering, transformation and loading data into data staging area within extract, transform and load (ETL) management processes [2].

Specific data from each information service (data source) can be masked and propagated into the appropriate data marts, which are subsets of the data in RAC data warehouse. The data marts contain aggregated (summary) data from heterogeneous information services (e.g. Oracle database, MS SQL Server,

MySQL, Excel) at the particular level of hierarchy. In this way, it provides more effective data structuring and eliminates the need to aggregate data when executing a query or analysis by end users. This leads to better performance and avoids the redundancy of data.

Data warehouse is a central source of consolidated and transformed current and historical data used by various professionals (analysts) for business analysis, data exploration and decision support [2]. This database can be accessed directly from an application level, which consists of GIS analytical machine (thick client) and the Report Server (BI Server). They are used by RAC analysts who have access to all data in read-only mode in order to implement the analytical operations.

GIS analytical machine is a generic set of analytical tools integrated through dedicated applications that are customized in terms of localization, workflow and typical data sets used for analysis. Also, in addition to direct access and read data from the data warehouse, they allow the entry of data which come from other sources, e.g. news media (Open Source Data) through the available forms. GIS tools provides spatial data analysis that can be conducted in order to indicate and analyse of spatial trends [7].

The analysis results are presented in corresponding form by the report generator (BI Server). Reports are available to all analysts and other authorized users of the system, and the representation of spatial phenomena can be achieved by using integrated components of spatial engine, application server and viewer (e.g. Oracle Spatial, WebLogic, MapViewer) consolidating results with a set of base geospatial data in the background.

4   Case Study: Location Intelligence System for Risk Analysis Center

4.1   The Aim of the Project

Building a functional and efficient integrated system for border management at the national and international level implies the establishment of the Center for Risk Analysis as its key parts. Bosnia and Herzegovina conducts activities related to these issues along with other duties essential to the process of liberalization of visa regime with the European Union. The concept of Integrated Border Management (IBM) involves coordination and cooperation of all state agencies and bodies involved in cross-border activities, in order to ensure maximum efficiency and effectiveness of border management.

The general objective is to create a safe and reliable system framework for risk assessment and spatial risk analysis which includes establishment of: basic functionality of the Center for Risk Analysis (CAR), communication channels and protocols for data exchange with professional services, or information services (IS) which supply the center with the source data, and functional information services or agencies and bodies for the collection, primary data processing and propagating it towards the CAR.

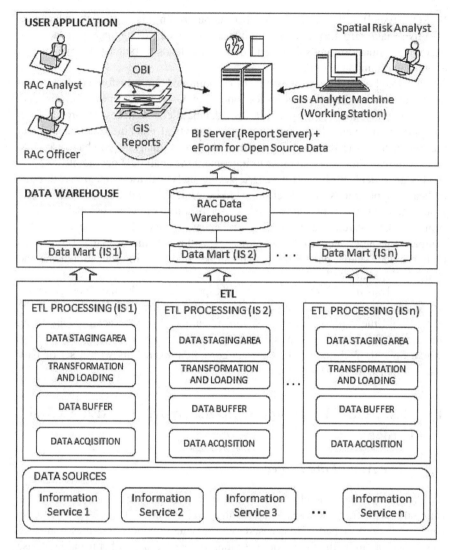

Fig. 1. Architecture of the business intelligence system with spatial extension for RAC

4.2   The System Implementation

System design is based on a central database to which data are propagated from information sources responsible for collecting, selecting and masking (protected) data relevant for risk analysis (Fig. 1). These data are prepared and transformed in a way that enable the production of quality reports of potential threats and risks that could endanger human lives, material goods or social order. Information Services (IS) are the various state institutions, agencies, professional services or local centers for risk analysis with its own organizational, informational, technological and technical characteristics. Integration of the spatial risk analysis system is directly related and dependent on the functionality of individual IS components. It is implemented through the phases of defining the scope, development, testing and deployment of this system.

The phase of defining the scope includes the following activities: information collection, analysis of the existing state and system design. During this phase, the final architecture is defined as logical and physical foundation on which to build the entire system of risk analysis. One of the major goals at this stage is surely evaluation of the integration of agencies (information services) in the proposed architecture (Fig. 1), in terms of compatibility of the existing systems with the future architecture considering equipment, databases, communications, and the specific needs of each agency.

During this phase, it is developed logical and physical architecture of the business intelligence system (BIS) as a configuration map data sources needed to finally form a single repository. After defining the logical configuration, it is defined the data flow from the source to the application level, as well as the general data access. Part of data propagated towards the center for risk analysis are masked by using special algorithms for encryption. Decryption key for unmasking the data should have only agencies that own ID data for conducting operational tasks.

Therefore, based on analysis of existing systems in some agencies with the specification of the available equipment, and other verified operational resources, as well as their operational and analytical needs is defined:

- logical system architecture,
- ata flow from the source to the user level,
- data sources (existing systems) through relational diagrams,
- the way of masking certain protected data before sending it to the CAR and
- other key elements for building the system.

Finally, the detailed design of all procedures and processes for the project is documented at the end of the design phase of the system. In this sense, through the final design solution is necessary to consider and address the following questions:

- the amount of data in a central database,
- extract / transform / load (ETL) of the data,
- data warehouse (DWH) design / database / collection process and design data mart's including: defining the dimensional model with the presentation in the form of diagrams, the convention on the level of object names, relationships between objects, the logical level of metadata and define the factual and shared database tables,
- data manipulation: security, access, updating, refreshing, replication, editing, archiving, backup, disaster recovery of the data,
- system management,
- testing the system and training of operating personnel and
- move into production mode, and system maintenance.

The development phase includes activities of development all the components that make up a complete system for risk analysis, their documentation, development of test scenarios and test data and training materials preparation. The testing phase includes activities of checking the readiness for testing and conducting of the final tests.

The deployment phase will be conducted when:

- all necessary equipment, software and communication network is installed and ready for use, and when the test environment is set up,
- programs and scripts for retrieval, transformation, load and refresh the original data in DWH are developed and tested individually in a test environment with appropriate examples of data,
- measuring loads in the process of loading data are executed,
- sample report, together with some ad-hoc queries are developed in the test environment and when the validity of results is verified and
- policy of access to data in the central database is established.

After the programs and scripts are developed and tested, it is performed the final inspection and testing of network resources and equipment, and overall functionality of the integrated system (before switching the system into production mode with real data). The transition from the test environment to production mode, means that production central database (data warehouse) is created and the processes of extracting, transforming and cleansing data is made in real systems with real data. At this stage, the development team and the operational staff conducts an initial uploading and procedure of the first data refreshing in the risk analysis system.

4.3 The System Using: Business Intelligence with Spatial Application in Risk Analysis of Epidemic Infectious Disease

For a number of analyses types conducted at the Center for Risk Analysis (e.g., trafficking, tracking shipments of plant origin...) there is a need for spatial presentation, or for the use of spatial analytical techniques. Therefore, the data propagated from the information services (or agencies) are geocoded and referenced in the spatial domain. Analytical capabilities directly dependent on available data, but also the level of detail displayed. At least, a phenomena or its trend can be displayed to the level of settlements, streets or border crossings. This is enabled by means of incorporated background (base) map data for the area of Bosnia and Herzegovina and the wider region. These data include: basic cartographic detail (cities, roads networks, ports, airports, railway network, administrative boundaries, border crossings) (Fig. 2), environmental data (climate, land use, precipitation, soil, forests, DTM, water bodies), utilities (power supply, telecommunication), demographic data and descriptive statistics (population, education, employment, agricultural yields...), economic data (administration bodies and development indices), and also Web services offering background geodata (satellite images, street maps).

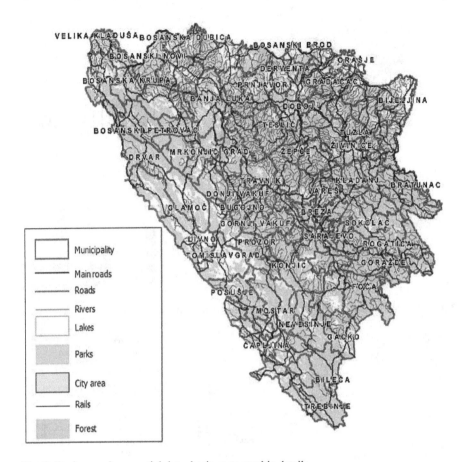

Fig. 2. Background geospatial data: basic cartographic details

User application environment, which includes the integration of GIS, Business Intelligence with Map Viewer (web application) (Fig. 3) covers a broad analytical functionality (geospatial analysis, business intelligence, reporting, publishing, creating a dashboard), which allows implementation of various types spatial data analysis from routine data presentation (eg. inspection of entries and exits at border crossings as shown in Fig. 4) to complex exploration techniques and concepts (geostatistics, clustering, gridding, network analysis, spatial queries, spatial data mining...).

The methodology of these analytical concepts is independent of the architecture and implementation of the system [8] that gives full freedom of analysts to create different models and scenarios during the analytical process. This allows the application of different methodological approaches that may include iterative stages such as framing the question, formulating the approach for addressing the problem, data acquisition, selecting appropriate methods and tools for analysis, and delivering the results and conclusions [9].

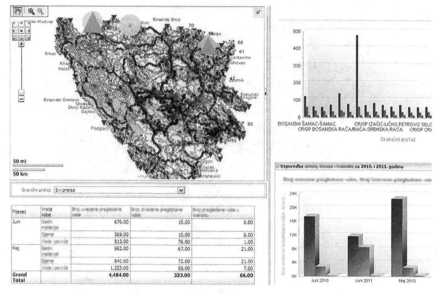

Fig. 3. Business intelligence dashboard for outlier detection

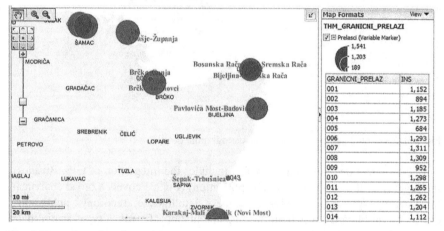

Fig. 4. Inspection of entries and exits at border crossings

One of the test scenarios used to conduct the analytical process and generate reports from the BI system is risk analysis of epidemic infectious disease to monitor the appearance of brucellosis caused by Brucella melitensis bacteria.

It is transmitted by ingesting infected food, direct contact with an infected animal, or inhalation of aerosols. Brucellosis primarily occurs through occupational exposure (e.g. exposure to sheep), but also by consumption of unpasteurized milk products.

This phenomenon is identified on the basis of information received from public health services. The task (set in this scenario) is to determine:

- spatial foci (hot spots) of disease,
- spatial trend of expansion,
- source and cause of the phenomenon
- measures to control epidemics and future prevention.

In short, this analytical process is conducted through five methodological steps [9].

The first step is the entering of external data (on the phenomenon with disease indications and number of patients) in the system. The phenomenon is registered at the locations (marked by symbol) that are found through address system search engine (based on the known address of infected individuals) (Fig. 5). Foci of disease are generated using the hot spot and cluster analysis.

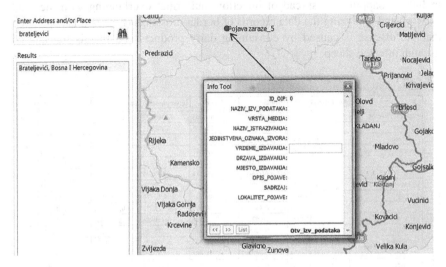

Fig. 5. Entering of external data with location of disease phenomenon

The second step is the selection of analytical method based on the entered data on the phenomenon and the available data sets in DWH (propagated from information services). The proposed method here is spatial autocorrelation, which examines the relationship between numerical grids. The matrix obtained by the method (Fig. 6) allows the investigation of correlation levels between groups of numerical grids (i.e. sets of attribute data from different information sources, e.g. records of foreigner residence registration, or tracking animals and plants shipments...). The elements of this matrix are the coefficients with values ranging from -1 to +1, where values close to 1 describes the attributes with a high degree of correlation (spatial dependence), and those that tend to 0 describe pairs of attributes with a low degree of correlation (spatial independence) [9].

The third step is the implementation of the analytical process, where the matrix analysis identified the following spatial relationships:

- there was increased presence of male foreigners with primary education (nomadic cattle breeder) and temporary residence nearby the hot spots for the previous period and
- there was migration of sheep herds nearby the sites of infection and a number of their border crossings.

The higher value of correlation coefficient does not imply a cause of the phenomenon, but it just says that there is a spatial relationship between certain spatial phenomena and the next task is to use other methods of analysis to determine the actual cause of infection [10].

The fourth step is spatial reasoning and inference on the cause of the phenomena [11]. On the basis of detailed exploratory spatial analysis which included: mapping the spread of infection, buffering, overlapping with the road network and analyzing the data records of border crossings (Fig. 7) it is concluded that the disease is caused by consuming dairy products and that is transmitted through infected sheep.

correlationResults Browser							
_#	AvgFamilyIncome	Dwellings	Employed	Unemployed	EdUnder9	EdHighSchool	Ed
AvgFamilyIncome	1	0,531563	0,604553	0,0386577	0,0806681	0,402644	
Dwellings	0,531563	1	0,830081	0,263635	0,363308	0,726107	
Employed	0,604553	0,830081	1	0,195487	0,282821	0,771115	
Unemployed	0,0386577	0,263635	0,195487	1	0,97934	0,177846	
EdUnder9	0,0806681	0,363308	0,282821	0,97934	1	0,253252	
EdHighSchool	0,402644	0,726107	0,771115	0,177846	0,253252	1	
EdUniversity	0,450548	0,792118	0,866593	0,177986	0,265125	0,797234	
SpkEng	0,592451	0,759092	0,892247	0,13739	0,212796	0,692397	
SpkFr	0,216211	0,454734	0,44909	-0,0622372	-0,0283771	0,562436	
SpkOther	0,154573	0,52135	0,427989	0,739864	0,856283	0,375379	
Population	0,566968	0,880658	0,951085	0,275766	0,37573	0,830333	
PopMarried	0,588297	0,73245	0,911409	0,174171	0,254365	0,795186	
PopSingle	0,436124	0,842002	0,748458	0,32493	0,420092	0,551616	
PopAge0_14	0,407566	0,630215	0,826932	0,258399	0,3436	0,753478	

Fig. 6. Results of spatial autocorrelation applied on data sets in DWH

The fifth step includes presentation of the results of the analysis and generation reports using GIS and BI Publisher tools.

5   Advantages and Benefits of the Approach

BI system for advanced analysis of threats and risks has been designed and implemented as a dynamic and flexible framework with a variety of possibilities for improvement and expansion in all segments. This solution will certainly contribute to the quality and consistency of IBM in Bosnia and Herzegovina.

Basic functionality and capabilities of the system can be further expanded and enriched by adding new data sources from existing agencies and other relevant institutions, by consolidation of existing transactional systems, regulation procedures and templates for making the required reports, analysis, etc. As well, this system can be expanded with special geoportal that would allow other institutions and agencies access to thematic spatial data from the data warehouse for risk analysis at the local level in other segments of risk assesment. The geoportal can allow direct access to raw data in multiple formats, complete metadata, online visualization tools so users can create maps with data in the portal, automated provenance linkages across users, datasets and created maps, commenting mechanisms to discuss data quality and interpretation, and sharing or exporting created maps in various formats. This empowers BI solution with complementary technologies including spatial ETL, data visualization, and geographic information systems [12].

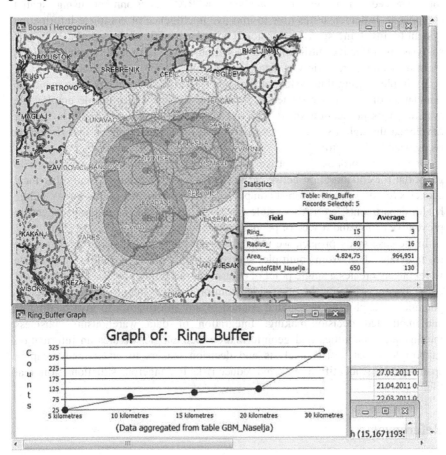

Fig. 7. Spatial reasoning: hot spot buffering and overlapping with the road network layer

For its implementation is a key issue to achieve high level of cooperation between RAC and individual information services (agencies). As well experience and knowledge of operating personnel (analysts and administrators) is very important. Lessons learned through design, implementation and use of this system can be extrapolated to other similar systems, with certain adjustments.

6 Conclusion

The implementation of a robust and complex BI system with a spatial extension, as a support to center for risk analysis, represents a challenge in any sense. Despite the availability of technological capabilities this requires additional innovation to achieve the goal and develop a functional and operational system. This paper shows an innovative approach for implementation business intelligence systems for advanced threat and risk analysis in RAC environment using spatial component. It demonstrates how to improve intelligence of complete information system by involving spatial extension. Particular benefit of this approach for risk analysis is effective utilization of location data, advanced spatial analysis techniques and more variety in data visualization.

By implementing this system at the center for risk analysis has been made the foundation of which is reflected in a modern, flexible and multiple employable system. In its production stage, it is not a closed system, but should be further developed through a series of possible improvements and options that should be recognized by its customers.

One of the proposed improvements is the extension of the system with geoportal for collaboration between the Center for Risk Analysis, and other local agencies (civil protection, forest companies, fire departments, police forces, public health departments, etc.) responsible for assessing the risks and threats to ensure preventive action and rescue of people and property.

In the case study project, we utilized standard data warehousing infrastructure to integrate data from multiple source systems and geospatial data. Data warehousing creates a single point of control for managing processes that cross the entire system, while on the other side geoinformation system (GIS) tools have the potential of exploiting the spatial context of any information to give qualitative and motivated decision-making. Integration of data warehousing, business intelligence capabilities and geoinformation technologies creates an information support tool that assists analysts and decision makers to understand complex spatial patterns, identify threats and reduce risks in dangerous situations from real life.

References

1. Gupta, J.N.D., Sharma, S.: Intelligent Enterprises of the 21st Century. IGI Global (2004)
2. Moss, L.T., Atre, S.: Business intelligence roadmap: the complete project lifecycle for decision-support applications. Addison-Wesley (2003)
3. Curtis, G., Cobham, D.: Business Information Systems: Analysis, Design and Practice. Pearson Education (2008)

4. Rigaux, P., Scholl, M., Voisard, A.: Spatial Databases - With Application to GIS. Morgan Kaufmann (2002)
5. Shekhar, S., Chawla, S.: Spatial Databases: A Tour. Prentice Hall (2003)
6. Karabegovic, A., Ponjavic, M.: Integration and Interoperability of Spatial Data in Spatial Decision Support System Environment. In: MIPRO IEEE Croatia Conference, Opatija, Croatia (2010)
7. Batty, M., Longley, P.A.: Advanced Spatial Analysis: The CASA Book of GIS. ESRI Press (2003)
8. Worboys, M.F., Duckham, M.: GIS: A computing perspective. Taylor and Francis (2004)
9. de Smith, M., Goodchild, M., Longley, P.: Geospatial Analysis: A Comprenhensive Guide to Principles, Techniques and Software Tools. Matador (2010)
10. Haining, R.: Spatial Data Analysis: Theory and Practice. Cambridge University Press (2003)
11. Burrough, P.A., Donnell, M.C.: Principles of Geographical Information Systems. Oxford University Press (1998)
12. Karabegovic, A., Ponjavic, M.: Geoportal as Decision Support System with Spatial Data Warehouse. In: Federated Conference on Computer Science and Information Systems, ABICT 2012, Wraclaw, Poland, pp. 943–946 (2012)

Model Driven Architecture and Classification of Business Rules Modelling Languages

Bartłomiej Gaweł and Iwona Skalna

Abstract. An organisation's activity under dynamic changes of business processes requires continuous improvement of business practices. This implies the necessity of refining decision making processes. Business rules [6], [8] enable experts to transfer enterprise strategy onto the operational level using simple sentences which, in turn, can automate reactions to subsequent events both inside and outside an organisation. The main advantage of the business rules is their simplicity and flexibility so they can be easily utilised by different organisations for different purposes. In order to represent knowledge in a pseudo-natural language understandable to information systems (business rules engines), notation and description standards are required. This study presents an overview of the most popular business rules description languages and proposes criteria for their classification. Based on those criteria a comprehensive classification of business rules modelling languages are provided.

1 Introduction

The central idea behind *Business Rules Engines* (BRE) or *Business Rules Management Systems* (BRMS) is that any organisation uses a decision logic to maintain operational and managerial tasks. The major objective of modern business rules management systems is to allow for business and technological processes to be described independently of the software system to be implemented. Nowadays, it is possible to automate business processes by implementing appropriate ERP, CRM, SCM or WFM solution. However, there are still some processes and decisions which are non-routine or cannot be well described by appropriate semantic language.

Rules play an important role among the most frequently used knowledge representation techniques. They may be used to describe different aspects of business. The biggest advantage of BRMS approach is that business rules are understandable for human and may be also directly used in software programs [4]. However, software programs and humans need rules to be expressed in different languages in order to understand them.

Bartłomiej Gaweł · Iwona Skalna
AGH University of Science and Technology, Al. Mickiewicza 30, 30-059 Krakow, Poland
e-mail: {bgawel,iskalna}@zarz.agh.edu.pl

M. Mach-Król and T. Pełech-Pilichowski (eds.), *Advances in Business ICT*,   123
Advances in Intelligent Systems and Computing 257,
DOI: 10.1007/978-3-319-03677-9_8, © Springer International Publishing Switzerland 2014

Different target audience and broad field of potential applications cause that there are a number of languages and approaches for modelling of business rules. In what follows, some criteria which may be used to select business rules modelling language being the most suitable for specific problem are presented.

2   Business Rules and Semantics Web

The definition of a business rule, coming from GUIDE [11], states that a business rule is *"a statement that defines or constraints some aspect of the business. It is intended to assert business structure or to control or influence the behaviour of the business."* In [10], Ross describes several basic principles of business rules approach. He believes that a language has the biggest impact on business rules expressiveness. Therefore, in the remaining part of this study business rules description languages are presented and classified.

Business rules are usually written in a form of *"if... then ..."* statement. The most important aspects of business rules is that they should be unambiguous. It means that all terms used in rules should be defined in a business vocabulary which defines all important concepts in a particular area of business interest. Besides the definitions of concepts, such vocabulary also defines relations between concepts. A formal description of rules and a vocabulary should be different for different target users.

From the managerial point of view, a goal of business rules approach is to improve the communication between humans. Hence, business rules and business vocabularies should be described in a language close to the natural one. Business rules management systems need a definition of a vocabulary and rules in formal language to avoid ambiguity of meaning. Nowadays, especially in supply chains, business rules managements systems participate in larger networks and exchange business rules. In such a case, an information context should be transferred by a language together with rules and vocabulary. Unfortunately, there are no languages which meets all those requirements. To solve that problem, two models - *Model Driven Architecture*[1] (MDA) and *Semantics Models* (SM) have been developed.

MDA model defines three abstraction levels. *Computation Independent Model* (CIM) shows how the system works within an environment, without details on the system's structure and application implementation. At this level, rules are written in a language with the syntax close to the natural one or are presented as diagrams (visual languages). *Platform Independent Model* (PIM) is a model of a software system or a business system that is independent of the specific technological platform used to implement it. At this level, rules have standard representation based on an XML language. Finally, at the *Platform Specific Model* (PSM) rules are described in a language designed for a specific BRMS. The idea of SM focuses on exchanging data between systems without human support. Hence,

---

[1] A detailed description of the MDA architecture can be found at
http://www.omg.org/mda/

Semantic Models develop languages which transfer both data and context between those systems.

MDA and SM concepts assume that there should be many very limited purpose languages and the appropriate translation programmes. The next section describes the differences between rules description languages.

3   Classification of Business Rule Modelling Languages

On the basis of the subject literature ([1], [2], [3], [9], [12]), the following set of the most popular modelling languages have been selected for the comparison purposes: the Semantics of Business Vocabulary and Business Rules (SBVR), the REWERSE Rule Markup Language (R2ML), the Semantic Web Rules Language (SWRL), the Rule Markup Language (RuleML), the Rule Interchange Format (RIF), the Java Specification Request (JSR-94), Prova and the Production Rule Representation (PRR).

First, the comparison of business rules description languages is performed at the respective abstraction level of the MDA architecture and the selected languages are assigned to those levels:

- Business models level CIM: SVBR language,
- Business functionality level PIM: PRR, R2ML, SWRL, RuleML, RIF languages,
- PSM level: JSR-94 language.

One can see that SVBR is the only standard that exists at the CIM level. It enables rules to be described in a language similar to a natural language.

At the PIM level there are five languages with different expression power. Thus, at the PIM level there is no one universal language which means that the language must be selected appropriately to specific rule type and vocabulary. RuleML is the only language that supports modelling of any rule. Strict assignment of languages to an appropriate type of rule is presented later in this section. Here, it should be underlined that all business rules description languages at the PIM level are written down using XML.

Languages at the PIM level are then transformed into languages at the PSM level. At this level rules modelling languages are strictly connected with programming languages used for developing specific BRMS. This is due to the fact that at this level, rules should be written in a way which guarantees computational effectiveness. Most of the BRMS (e.g., Drools, ILog) is written in Java. Hence, JSR-94 is a standard at the PSM level as an API Java library.

Next, a comparison of meta-languages used to write down standards of business rule modelling were made. The SBVR model is written using a natural (English) language, and dependencies between themes are written using the XML and SBVR XML Schema. Two meta-languages can be distinguished at the PIM level. Rules are written in XML, and respective description schemes in MOF/UML (PRR, R2ML, RuleML), or EBNF formalism (RuleML, RIF, SWRL).

At the PSM level, languages are described in a way which guarantees their compatibility with the JSR-94 standard or Java.

Another comparison checks how business rules modelling languages *integrate with external ontologies or databases*. The integrations are usually implemented through establishing links to dictionaries contained in ontologies. These links are most often implemented through URI or IRI. Common usage of URL (*Uniform Resource Locator*) in a semantic model dates back to the work by Heflin and Hendler [7].

A dictionary is seen as an external data type and is represented by external objects. Owing to this solution, rules description languages retain quite a simple construction. Table 1 shows the comparison of usage of business rule modelling languages.

Table 1. A comparison of languages used to link to external fact bases

Language	Standard identifier	
	URI	IRI
R2ML	X	
RuleML	X	X
SQL	X	
Jena Rules	X	X
OCL	X	
RIF	X	X
PROVA	X	

In most of the markup based languages (R2ML, RuleML, Jena Rules, OCL, RIF, SWRL), URI identifiers are used to identify facts. Some of them, such as SWRL, have special markups (e.g., `swrlb:resolveURI`) which enable operations on arguments described by URI. All classes described by R2ML have also URI references. Class is a type for R2ML objects or variables. Similarly, reference property and datatype predicate in R2ML have URI references. RIF uses IRI as constants, similar to RDF.

The comparison described above clearly shows that there is one standard, SBVR, which allows the production of business vocabularies and rules to be written in a language close to the natural one. This standard is dedicated to human audience. The same role for computer audience plays JSR-94 closely related with Java. Both of these languages may be easily translated to PIM level languages. Because of different expressiveness of PIM level languages, the choice of the inappropriate intermediate translation language may result in incomplete translation or translation lost. Therefore, the following part of the paper focuses on the following languages from the PIM level: R2ML, SWRL, RuleML, RIF.

4 Comparison of Structures of PIM Level Modelling Languages

A knowledge database is a structure which consists of several levels. For the PIM level languages comparison purposes, the three among the latter are the most important: data level, rules level and knowledge representation level. At the data level, *methods for building particular languages* should be compared. Scheme type and validation are reliable comparison criteria.

Dependencies between structures in business rules modelling languages are usually described using XML Scheme. There are two approaches to build the latter. A language can be described by a flat scheme or by a structure in which each construct is described by a separate module, which is then used to build ontology. The second hierarchical approach, unlike flat structures, can easily be extended. Moreover:

- a programmer needs to use necessary modules only (e.g., those adjusted to some type of rules), he does not have to use all elements,
- it is much easier to develop a language described by a modular scheme,
- modular approach facilitates testing of the solution used to implement the model.

R2ML are based on flat scheme. On the other hand RuleML, SWRL, and RIF follow the second approach. They define respective constructs in separate modules.

Rules are built on facts and terms which creates business vocabulary. A business term is a word or phrase that has a specific meaning for a business in some designated context. A fact asserts an association between two or more terms. That is, it expresses a relationship between the terms. Business rules description languages on PIM level enable a flexible facts notation. Fact can be represented by: term, class, object instance, state or event. A fact can be defined directly in a language or can be assigned to a rule through a proper addressing language URI and IRI (described above).

One of the most important question concerning directly defined facts is the question of whether facts and terms are stored as attributes or elements. If business rule contains information which can be validated, then an element should be used. Otherwise, one can use attributes.

R2ML differs from other languages at the PIM level, because it writes all information items as attributes in contrary to other languages which uses elements as well. However, all of the PIM level languages enable the facts to be described as simple and complex datatypes. For example, beside simple representation of an attribute, RuleML introduces possibility of organising attributes in multisets (`arg`) and sequences (`slot`) using the index attribute. Similar notation is used by SWRL and RIF. R2ML does not distinguish between attribute and element notation.

Direct defined facts are described in different ways in different languages. RuleML provides a certain set of constructs which can be used in any semantically significant combination. Thanks to this, RuleML rules are easy to write and maintain, and can be automatically translated into other languages. SWRL and RIF, which are based on R2ML, introduce more complex structures.

In contrary to RuleML, R2ML very strongly differentiates a type of terms and atoms. Owing to this, R2ML has a very rich and extensive syntax. For example, a variable in R2ML can be represented by an variable responsible for objects, literals, and (variable without a type). In RuleML and its derivatives, there are no such division. To distinguish between an object and a value, R2ML, similarly to RuleML, introduces respective objects for an object name and a value of a variable.

Languages for writing rules in business rule engines are usually based on a Prolog (Prova, TRIPLE, or Jena) syntax or on their own syntax adjusted to a programming language used to write the engine.

There is a wide range of business rules applications. Therefore, different type of rules can be distinguished [12]:

- *integrity rules* are used to express conditions that should be fulfilled, e.g., *"Each new person driving a staff car must be a qualified driver"*.
- *derivations rules* describe how data item can be computed from other data, e.g., *"A customer is considered as a premium customer if it spends more than USD 3,000 a year"*.
- *reaction rules* respond to an event occurrence, e.g., *"If a customer has returned a rental car and the mileage is more than 8,000 kilometres since last servicing, then send the car to the service"*.
- *production rules*, in contrary to reaction rules, they are not called by any previous event, they "produce" events if the conditions are fulfilled. An example of the production rule at the CIM level could be: *"If order value is greater than 2,000 and the type of customer is not Premium, then a 5% discount is granted"*.
- *transformation rules* describe constraints which must be fulfilled to change operational data format, e.g, *"If you receive a full description of the book, then transform it to a short description containing the author name and book title only"*.

Realisation of particular types of rules is possible through the introduction of what is known as procedural attachments and built-in functions. A business rule must often recall queries and procedures stored externally and call external sources of data. Such procedures, called procedural attachments, can be used in business rules.

Table 2. Comparison of how particular languages support occurrences of subsequent rules

Language	Rule type				
	Integrity rule	Derivation rule	Reaction rule	Production rule	Transformation rule
R2ML	X	X	X	X	
RuleML	X	X	X	X	X
SQL	X	X	X		
Jena Rules	X	X	X		X
OCL	X	X	X(OCL)	X(OCL)	
RIF	X	X			
SWRL	X	X			
PROVA		X	X		

A procedural attachment is a function or a predicate which is implemented externally. There are two types of such attachments:

- logical attachment, which returns `true` or `false`,
- object, which gets certain objects and returns one or more objects.

In addition to procedural attachments, the built-in function can be mentioned. They are functions built in rule description languages, or predicates which enable operations to be performed on, e.g., strings, numbers or logical values.

SWRL and RuleML have the common library of built-in functions, which is based on the functions available in XQuery and XPath languages. Built-in functions are called in SWRL by a special atom `swrlx:builtinAtom`.

SWRL does not support procedural attachments, but they might be called directly from ReactionRuleML. R2ML supports built-in functions by default, in the same way as SWRL does, but uses `r2ml:DatatypePredicateAtom` predicates and `r2ml:DatatypeFunctionTerm` functions. Besides the function, an `r2ml:DatatypePredicateAtom` element contains variable name and type. Thanks to this, in the next step, it can be projected on the respective operator with the respective data.

R2ML enables the introduction of procedural add-ins that, as depicted in the listing, allows access to any external function.

To represent production rules, R2ML was enhanced with constructions which enable action-reaction rules to be described.

Knowledge representation is here understood as a way of presenting knowledge about the world together with processing, especially inference, methods. From the business rules' point of view, the following knowledge representations can be distinguished [5]:

- *decision table* – represents knowledge in a form of a table, which contains prerequisites and conclusions. This kind of knowledge description supports the creation of models containing numerous independent conditions,

- *decision grid* – supports the presentation of rules which are functions of two or more related conditions,
- *decision tree* – presents rules as a tree-like graph which shows decision process. In the case of numerous tied rules, a decision tree clearly shows the history of decisions. The main deficiency of decision trees is that they do not support the application of advanced solutions, such as AND or XOR, for controlling processes,
- *scenarios* – describe stereotypical sequences of process events using special scenes corresponding to possible decision situations.
- *workflows* – represents rules as a set of actions that are executed when certain specified conditions are met. Workflows were adopted for presentation of knowledge using tools for description of system workflow, a timing of system events can be presented.

All described standards enable rules to be combined in the form of a decision table. Subsequent rules are written one under another in the form of an XML structure. Rules cannot be written in the form of a workflow in any of the standards. Rules can be written in the form of a workflow using the token language BPEL, which supports combining decision rules.

5   Conclusions

The most important business rule description standards are presented in this study. Selected from the recent literature, the most common business rule modelling languages, such as SBVR, R2ML, SWRL, RuleML, RIF, JSR-94, Prova, PPR, have been characterised and compared. In the first step, they were grouped to account for different goals they realise and to make it possible to compare them. The MDA architecture and Semantic Web were selected as classification criteria to create groups of comparable languages. Based on the MDA classification, the following conclusions can be drawn:

- First of all, it should be noticed that there are no useful business rule description standards at the CIM level,
- At the PSM level there exist one standard only for communication with business rules engines - JSR-94. It should be emphasised that it is compatible with the Java language,
- There are no languages supporting efficient business rule visualisation
- There is a lot of languages at the PIM level; however, none of them can be translated to SVBR,
- Most of the standards do not support writing all types of business rules - RuleML is an exception,
- There are no languages based on SQL,
- It is difficult to draw conclusions based on the XML rule notation. The SQL notation would be more accurate.

References

[1] Antoniou, G., Damasio, C., Grosof, B., Horrocks, I., Kifer, M., Maluszynski, J., Patel-Schneider, P.: Combining Rules and Ontologies - A Survey. Deliverables I3-D3, REWERSE (March 2005), http://rewerse.net/deliverables/m12/i3-d3.pdf

[2] Boley, H.: Are Your Rules Online? Four Web Rule Essentials. In: Paschke, A., Biletskiy, Y. (eds.) RuleML 2007. LNCS, vol. 4824, pp. 7–24. Springer, Heidelberg (2007)

[3] Bry, F., Marchiori, M.: Ten Theses on Logic Languages for the Semantic Web. In: Fages, F., Soliman, S. (eds.) PPSWR 2005. LNCS, vol. 3703, pp. 42–49. Springer, Heidelberg (2005)

[4] Gottesdiener, E.: Business RULES show power, Promise. Issue of Application Development Trends 4(3) (1997)

[5] Graham, I.: Business Rules Management And Service Oriented Architecture: A Pattern Language. Wiley (2007)

[6] Halpin, T.: Business Rules and Object Role Modeling. Issue of Database Programming & Design 9(10), 66–72 (1996)

[7] Heflin, J., Hendler, J., Luke, S.: SHOE: A Knowledge Representation Language for Internet Applications. Technical Report CS-TR-4078, UMIACS TR-99-71 (1999)

[8] Morgan, T.: Business Rules and Information Systems. Addison-Wesley Publishing, Boston (2002)

[9] Paschke, A.: Rule-Based Service Level Agreements - Knowledge Representation for Automated e-Contract, SLA and Policy Management. IDEA Verlag GmbH, München (2006)

[10] Ross, R.: Principles of the Business Rule Approach. Addison-Wesley Information Technology Series (2003)

[11] The Business Rules Group. Defining Business Rules, What are they really? (July 2001), http://www.businessrulesgroup.org

[12] Wagner, G., Damasio, C., Antoniou, G.: Towards a general web rule language. International Journal of Web Engineering and Technology Archive 2, 181–206 (2005)

Author Index